Required citation:
FAO. 2022. *Fish – Know it, cook it, eat it*. Rome. https://doi.org/10.4060/cc1395en

Fish

KNOW IT, COOK IT, EAT IT

Food and Agriculture Organization of the United Nations
Rome, 2022

Fish
Know it, cook it, eat it

Author: André Vornic

Editorial adviser: Audun Lem
Editorial coordinator: Suzanne Lapstun
Copy-editor: Amy Knauff

Art direction: Monica Umena
Layout support: Simone Morini
Illustrations: Nik Neves

FAO technical inputs:
Molly Ahern, Fabio Consalez, Giulia Loi, Pierre Maudoux, Raffaella Rucci,
Jogeir Toppe, Rumyana Uzunova, Stefania Vannuccini, Weiwei Wan, Yimin Ye

Research and production support: Daniela De Pascalis
Production support: Laura Galeotti

CONTENTS

CONTENTS

CONTENTS

Great chefs cooking great fish

Ángel León
Fish mortadella
196

Malek Labidi
Salt-baked seabass in sauce vierge
198

Megha Kohli
Raw mango fish curry
200

Rodrigo Pacheco
Prawn nea piaraka soup with cassava crisp
202

Sandro Serva Maurizio Serva
Eel with a watercress infusion and kiwi
204

FISH, FAO AND YOU

COOKING WITH CONTEXT

Eating fish is one of the great pleasures of life. This book stems from that notion: it contains dozens of recipes from all principal regions of the planet. If you like seafood, aquatic foods and fishery products, there'll be something in here for you. At the Food and Agriculture Organization of the United Nations (FAO), with a presence in two-thirds of all countries, we know what ends up on the world's plates. The urge to share some of that culinary wealth was reason enough to bring this volume to life. But it wasn't the only one.

Most of the dishes we present are "national" or folk recipes, plucked from the gastro-lore that is the sum of household experience. The majority have been collected locally by our staff; often, it's fare they grew up with, or versions of it. Yet you'll also find a selection of dishes magicked up for us by celebrated chefs, in whose hands the national tradition is reimagined.

For each species of fish featured, we have sought to provide more context than cookery books normally offer – cultural, historic or environmental, and touching on matters of economics and trade. Food in our times is, after all, a form of world-wise awareness as much as a question of mere sustenance. And while most cookbooks start with the fish sitting on your kitchen worktop, ours starts at the fish stall – and even underwater – with tips on gauging the merchandise and avoiding the common phenomenon that is fish fraud.

FISH AGAINST POVERTY, HUNGER AND MALNUTRITION

The point is this: we at FAO think eating fish is not just pleasurable, but the expression of a global civic consciousness. This book comes out in the International Year of Artisanal Fisheries and Aquaculture – our cue to celebrate the contributions of fisherfolk, fish farmers and fish workers; to support their training for efficiency gains; to connect them to markets – which would include you, our readers; and to help them not just survive, but thrive. Here and there throughout this book, we'll be telling you what FAO is doing on those fronts.

But there's an even broader point. Under the Sustainable Development Goals (SDGs), the world has committed to ending hunger and malnutrition in all its forms by 2030. Throw in resurgent conflict and instability, the effects of the COVID-19 pandemic and weak political will, and that deadline is very much in doubt. In 2021, up to 828 million people were undernourished. Three billion individuals couldn't afford the most modest healthy meal.

Against this background, fish provides a fifth of the daily animal protein intake for almost half the world's population – and an even higher proportion if we consider just high-quality animal protein, or the micronutrients needed for diversified diets. In other words, fish may not suffice to ensure global food security, but there'll be no global food security without fish. Every piece of fish on an adult's plate is a blow to global hunger. Every school meal that contains fish is a win against child malnutrition. To eat fish is not just to bolster livelihoods: it is to partake of one of the world's most vital resources.

THE URGENCY OF RESOURCE MANAGEMENT

But if you'll forgive us an oxymoron, resources must be mercilessly nurtured. Another SDG for 2030 urges us to conserve life below water and sustainably use marine resources. As things stand, a third of these are being harvested at biologically unsustainable levels. Large sections of our oceans and seas are poorly governed. Criminal practices endure.

Encouragingly, though, conservation has crept up the agenda. This year's UN Ocean Conference in Lisbon has been a chance to grapple with the dramatic loss of marine biodiversity caused by climate change. A treaty to end harmful fisheries subsidies, under the auspices of the World Trade Organization (WTO), is gathering momentum. Almost 100 nations have signed up to FAO's Agreement on Port State Measures (PSMA): the parties agree to deny access to, and use

Fish makes up a fifth of the animal protein intake for half the global population
Quang Nguyen Vinh on Pexels

of, their ports to vessels that engage in illegal, unreported and unregulated fishing. By joining the PSMA, least developed countries are unblocking European Union market access for their fish products.

A FUTURE OF FISH FARMING

Markets, in fact, are driving elements of progress. Over half the global demand for fish is now met from aquaculture. For all its faults, fish farming helps relieve pressure on the oceans; it is, overall, more efficient than land-based animal production systems; and it generates fewer emissions. It's also an area of constant innovation, from non-medicalized technologies against fish disease to the use of cultured insects in replacement of fishmeal. The world needs to farm more fish, not less – ideally a third more by 2030, with the emphasis on food-deficit regions, primarily Africa – but do so sustainably.

Logically enough, this book makes no bones about featuring farmed fish alongside fish from capture, and frequently recommends its use. In the case of salmon, for example, where public criticism of breeding conditions has at times been intense, we discuss things frankly, sort fact from hearsay, and set out to allay consumer concerns.

Inspections, here conducted jointly by FAO and Peru, help deter illegal fishing
© FAO/Ernesto Benavides

FINALLY...

... our competence at FAO may be rooted in the science and economics of food, but we're also awake to food's convivial and emotional dimensions. Weighty mandate aside, we enjoy a laugh or anecdote more than we do any sort of lecturing. Which is why this book, while serious, is anything but self-important. We hope you'll find pleasure in our "fish interviews": a light take on interspecies communication, they put a playful spin on the project, and so do our recipe illustrations. If Fish: Know it, cook it, eat it piques your curiosity, tickles your conscience, entertains you, and above all leaves you well fed, then our job is done.

WHICH **FISH** AND WHY

There are probably more aquatic species in the world than there is anything of anything (microbes excluded). About 250 000 of these species are currently known to us, though estimates suggest there may be eight times as many in existence. Projections made in the 2010s concluded that 91 percent of marine organisms had yet to be described. Naturally, only a handful of this diversity of beings are fish, crustaceans, fish-like organisms or even algae; those fit for human consumption are fewer still.

Even with these caveats, a cookery book – albeit one with a taxonomic touch – must involve some triage. In deciding which fish to feature, we have looked to balance the familiar and the curious; to mix the universal and the local, the wild and the farmed, the saltwater and the freshwater; and – as befits an international organization with nearly 200 Members – to paddle far and wide, leaving no main region uncovered.

Of course, not all species will be available in all places. In fact, no single place on earth is likely to feature the range of fish featured here: we're well aware that our pursuit of diversity will run into practical availability concerns. Further, in some cases, what can be sustainably obtained in one region might not be so elsewhere. This is why we have often suggested replacements – not only for fish, but also for accompanying vegetables and other ingredients – in the hope that readers will be inspired to think up equivalents themselves. In the kitchen, to coin an ad hoc phrase, substitution is the mother of empowerment.

Amberjack

SERIOLA DUMERILI

....

Amberjack is splendid, tasty and a little muddled. Or rather, muddled is what we are, when it comes to deciding what to call it. In Latin we're quick to identify it as *Seriola*, from the Carangidae family of fast-swimming fish that inhabit warmer oceanic waters. In our vernacular tongues, we seem unsure. In American English, the greater amberjack is alternatively known as yellow jack, but on occasion as pompano, which may also designate the whole of the jack genus. In French-speaking countries, it could be a *sériole couronnée* or a *sériole ambrée* – when it's not a *limon* or, endearingly, a *carangue amoureuse*. Formal Italian uses *ricciola*, but regional dialects cling to their *leccia*, *fijtula* or *jarrupe*.

Japan, which farms more amberjack than any other fish, combines precision and haziness in referring to its native species, the *Seriola quinqueradiata*. There's certainly a full lexicon tied to the age of the animal, with *hamachi* (young fish, weighing around 3 kilograms) and *buri* (mature fish, 5 kilograms and up) the names most commonly seen on menus and stalls.

And yet, what to Osakans is *hamachi* can be sold in Tokyo as *inada*; *buri*, especially when farmed, may turn up as *hamachi*; and all are rendered in English as yellowtail, at the risk of being mistaken for the entirely different fish that is yellowtail tuna. (For more on yellowtail tuna, see p. 142.)

The potential for consumer frustration, in other words, is great. Or perhaps we should just relax into this normative bouillabaisse and hail the amberjack in its wealth of guises. After all, the hotchpotch of names attached to the creature conveys a form of organic authenticity: it suggests amberjack was embedded in local fishing and food cultures long before speech and commerce were standardized. Today, amberjack has nothing like the global following that cod and tuna have. But it soon might: this is one fish that's both plentiful in the wild and easy to farm. It lives 15 years at most, but it breeds prolifically. The flesh is buttery, packed with protein, and rich in phosphorus and potassium – the kind of minerals on which our muscles, nerves and teeth thrive. Not least, amberjacks catch the eye: they can exceed 1.5 metres in length, the female outstripping the male. Shades of gold – and, clearly, amber – light up the fins. Or, depending on the species, a citron line may bisect the fish from eye to tail, like a sunbeam slicing the clouds.

In a pattern some of us mammals might recognize, amberjacks are sociable animals when young, but grow more solitary with age. The Japanese market tends to allocate the younger fish, raised in pens off the southern islands, for sushi. The mature fish are usually caught in the wild: richer and more sapid, they are frequently paired with teriyaki sauce. To make this dish, sprinkle salt over a couple of thick amberjack steaks and set them aside for five minutes: the salt will draw out the blood. Wipe the steaks off (but do not wash them), dust them with flour, then fry them in a pan with a little hot oil. Press down on the steaks with a spatula to maximize crispness. Once the fish is seared on both sides, remove it from the pan, soak up the excess grease with kitchen paper, and set aside. For the teriyaki sauce, add equal amounts of soy, sake, mirin rice wine and sugar to the pan and reduce to a liquid glaze. Return the fish to the pan and slow-cook for another minute, basting it with the glaze. The flour on the steaks should further thicken the sauce: dilute with a few drops of water if needed. Serve over rice, with charred leeks and peppers, alongside green tea or cold beer.

Buttery flesh, packed with minerals

KNOW
YOUR FISH

Unless you catch your amberjack in the open sea – and a fair number of people do: amberjack is a popular recreational game fish – you will come across pre-cut fillets, steaks, or sushi and sashimi pieces. In finer Italian restaurants, amberjack may be served as carpaccio, raw and sliced thin to the point of transparency, though it's rarely sold that way.

If imported from Japan, the fish will almost certainly be farmed. It should be labelled yellowtail (again, watch out for any mix-ups with yellowtail tuna), or Japanese amberjack, or some combination of these terms. Elsewhere, amberjack may be from capture. Either way, the label ought to make this clear. In the European Union, the FAO fishing zone must be specified. If you're not sure, ask – but don't let the answer sway you. Wild or farmed, it'll be good regardless.

Amberjack is an oily fish: look for flesh that is plump, firm and supple. It should be rose-coloured in younger individuals, possibly with a bluish tinge; in mature fish, the flesh may run to a warm mauve. If the cut comes from the back of the animal, the bloodline is sometimes sharply delineated, brick red against pink.

Try not to overcook amberjack; you'll want to keep some fat on the fish, or it might go stringy and lose flavour. Flash-fry it, grill it or braise it with enough liquid to retain moisture. Our featured recipe comes from Cabo Verde, off the coast of West Africa. It calls for steam-poaching the fish, known locally as *charuteiro*, into a scrumptious Creole chowder.

Nutrition facts

GREATER AMBERJACK, RAW
per 100 grams

ENERGY (kcal)	129
PROTEIN (g)	21
CALCIUM (Ca) (Mg)	15
IRON (Fe) (Mg)	0.6
ZINC (Zn) (Mg)	0.7
IODINE (I) (μg)	11
SELENIUM (Se) (μg)	29
VITAMIN A (RETINOL) (μg)	4
VITAMIN D3 (μg)	4
VITAMIN B12 (μg)	5.3
OMEGA-3 PUFAS (g)	1.07
EPA (g)	0.19
DHA (g)	0.73

What do they call you?
Officially, *Seriola dumerili*, Carangidae family. That's what it says on my mailbox, should you wish to write to me. Beyond that, I have many folk names that often reference good things, such as gold or love. But even within the same language, people struggle to agree on one.

Who named you formally?
Antoine Joseph (also known as Antonio Giuseppe) Risso, in 1810. He was a naturalist from Nice, who published, among other things, an ichthyological study of what we now know as the French Riviera. He gave me the *dumerili* name in honour of another French scientist of the era, André Marie Constant Duméril.

You say we might want to write to you. But I don't know where you live.
In tropical and subtropical (45° north–28° south) areas of the Atlantic and Indo-Pacific Oceans.

Refresh my geography, please!
From Nova Scotia to Brazil in the western Atlantic, and off South Africa, in the Persian Gulf and around Australia, Japan and Hawaii in the Pacific. Also in the Mediterranean, and from the Gulf of Biscay to Senegal. In UK waters, not so much. In Japan, I also live in pens.

So you have multiple addresses, as well as multiple names. Can you tell us anything more personal about yourself?
I'm gonochoric, which means I split into male and female. This tends to happen when I'm 4 or 5 months old and about 25 centimetres long. But there's no sexual dimorphism: I don't look different if I'm one sex or the other. In that, I differ from other fish such as salmon – not to mention most humans, or lions and lionesses, or mandarin ducks.

THE INTERVIEW
AMBERJACK

Amberjack chowder

CALDO DE PEIXE

In one generation, Cabo Verde has leapt from least developed to middle-income country status. But its economy remains fragile. Just over half a million people live on nine scattered islands; a drought or a pandemic such as COVID-19 can quickly tip households into poverty and threaten access to food. Collective memory is still scarred by the pre-independence *Desastre da Assistência*, or Welfare Disaster, which struck as famine loomed in 1949. On that February day, thousands of Cabo Verdeans had gathered at the welfare office in the capital, Praia. They were queuing for food aid when a wall collapsed on top of them. More than 230 were killed.

Closer to our times, Cabo Verde's seafood sector acts as a social safety net and export earner. Paradoxically, it's proving harder to persuade the local hotel sector – the country draws much of its revenue from tourism – to source its fish domestically. FAO has meanwhile been helping Cabo Verde develop sustainable fishing practices and manage its

marine ecosystem: at the time of writing, the *Dr Fridtjof Nansen*, a research ship operated by FAO in a Norwegian-funded programme, had just set off on its third mission to the islands.

The fish trade in Cabo Verde is a gendered business. The men

catch and the women sell, either at markets or door to door. The cuisine, in this oceanic nation with ten times more water than land area, has much to offer: it draws on the same blend of Afro-Portuguese and tropical sensibilities that bubble through Cabo Verde's famed music scene.

TO FEED **4 TO 6 PEOPLE** YOU WILL NEED

2 LARGE GREEN BANANAS, OR 3 SMALLER ONES

800 G AMBERJACK FILLET

800 G CASSAVA (MANIOC)*
root, peeled, sliced 1.5-cm thick, cleaned and presoaked

½ BUTTERNUT SQUASH
(or acorn squash, pumpkin, etc.), peeled and diced, about 300 g

1 KG YAMS OR SWEET POTATOES, SLICED

2 CAPSICUM PEPPERS,
of different colours if available, diced

2 OR 3 LARGE TOMATOES, PEELED AND CHOPPED
(To peel the tomatoes more easily, score the skin and blanch them for a minute in boiling water, then leave them to cool. The skin will curl up and separate.)

½ HEAD KALE
(leaf cabbage), chopped into 3-cm sections

1 COMBINED HANDFUL PARSLEY AND CHIVES, CHOPPED

2 ONIONS, SLICED

6-8 CLOVES GARLIC

1 BAY LEAF
•
SALT AND FRESH PEPPER
•
OLIVE OIL
(alternatively, half olive and half sesame oil) and vinegar

*For a wider discussion of cassava, see River fish of the Amazon on **p. 160**. Note that you should start soaking your cassava slices well in advance to eliminate toxicity.

METHOD

1 In a mortar, make a paste with half of the garlic, salt and pepper, 2 tbsp of oil, the bay leaf and a splash of vinegar. Rub this paste all over the fish fillets and set aside in the fridge for an hour.

2 While the fish is marinating, peel the bananas and cook them in salted boiling water for 20 minutes or so.

3 In a large pan, gently heat more olive oil. Crush the remaining garlic roughly and add it to the pan to soften. Add a layer of onion slices, then a layer of tomatoes and herbs. Place the yams, capsicums and kale on top, then top off with the remaining onions, tomatoes and herbs. The tomatoes should moisten the mix, but you can add a little water if needed. Continue to sweat the lot, taking care not to burn it.

4 Now place the fish on top of the vegetable mix, season the lot generously with salt and pepper, and pour in the lime juice. Add enough water to cover the vegetables, but not the fillets – the fish will cook by absorbing the steam. When the water has come to a boil, turn down the heat and simmer. Replenish the water if necessary, keeping it below the fish line.

5 When the vegetables look nearly done – always best to taste! – add the diced squash and the pre-cooked bananas cut into thick slices. Pour the coconut milk onto the mix and bring to a boil.

6 Turn off the heat. Carefully remove the fish fillets to a side dish. Ladle the vegetable chowder into individual bowls, place the fish on top, and serve.

Anchovy

ENGRAULIS ENCRASICOLUS
(also ~ ANCHOITA, MORDAX, JAPONICUS, RINGENS, CAPENSIS)

....

Anchovies (Engraulidae) are found in many of the world's temperate zones. From Worcestershire sauce to *nước mắm*, places as different as England and Viet Nam have a history of bashing, mashing and squeezing them into some form of flavour enhancer. That aside, economically as much as gastronomically, the anchovy is foremost a fish of the Mediterranean and its cultural catchment areas: the Iberian Atlantic and the Black Sea. You don't need special map-reading skills to glimpse the backbone of the Roman Empire through today's main anchovy production areas. Italy, Spain, Portugal and Türkiye vie for top spot; Greece and Croatia line up behind them.

Silky *filetes de anchoa* of the Cantabrian Sea, off northern Spain, packed in collectable period tins. *Colatura di alici* from Italy's Amalfi coast, heir to the prized Latin extract garum, kept alive through the centuries by Cistercian monks. An anchovy swirl over a dollop of burrata with a side of puntarelle, the curly Roman salad, dressed in an anchovy vinaigrette – all of it then tossed over sourdough pizza. A couple of cured fillets, stirred into olive oil on the hob until they melt to bequeath a warm savouriness, in the way of a scent after its wearer has left. Or a Neapolitan street snack, consisting of a pile of the fish – tiny, whole but for the heads, and fresh rather than filleted and cured – rolled in flour, deep-fried and served in greaseproof paper. The anchovy is ancient and modern; swish and homely; it is, like ripe cheese, both jolting and refined. Salty, no doubt – but no more so than a bag of crisps, and much more nutritious to boot. Anchovy is rich in niacin, an essential B vitamin, and in selenium, which protects against cell damage.

The analogy with cheese is not gratuitous. Anchovy *does* go with dairy, in a way that a lot of fish doesn't. This is not to say that haddock, for example, isn't good steeped in milk, or that canned tuna won't work well topped with grated cheddar and shoved under the grill. Still, the tiny, feisty anchovies win the versatility contest. Pair them with butter, with cream, with gooey stracciatella or slivers of manchego; with garlic, vinegar, olives and chillies; with pale foils or fiery equals – all make good partners.

Strictly speaking, anchovies fall into well over 100 species, though only six or so register commercially. Differences can be minute: anchovies are universally small and slender; silver in colour, with an occasional greenish tinge; and closely related, in both looks and genetics, to the larger herrings, sardines and pilchards. (Herrings are specifically discussed on p. 80).

Immensely gregarious and, given their diminutive size, low on the marine food chain (almost everyone preys on them), anchovies move in schools so vast and tight as to darken entire portions of the sea. One school that snaked for three days off the beach in La Jolla, California in July 2014 may have been the largest ever observed: untold millions of individuals appeared to form a giant shadow crossing the water's surface. The school's presence so close to shore, and in waters so warm, baffled scientists – and it still does. We may, in other words, have been eating anchovies for thousands of years, but we've yet to fully unpack their cryptic allure.

Low on the food chain, high on mystery

KNOW
YOUR FISH

Whether whole, filleted, preserved in brine or vinegar, or cured in salt and nestling in olive oil, you will often buy your anchovies in a tin or a jar. Castilian Spanish, in fact, has three separate names for the anchovy depending on its state: *bocarte* when fresh, *anchoa* when salt-cured and *boquerón* when preserved in vinegar. Prices rise in line with geographic origin – a Cantábrico or Cetara label, from Spain and Italy respectively, will add to the expense – but also with the amount of labour and skill deployed to fillet the fish. Anchovies that are hand-processed (it will say so on the tin) and packed within hours of being caught will set you back the most.

If getting your anchovies fresh, expect to find them 10–15 centimetres long. Some countries set a minimum legal length: in Türkiye, where our recipe comes from, this is 9 centimetres. In early 2021, when Black Sea catches started returning shorter-than-usual fish – a combination of poaching and climate change was blamed – the Turkish authorities instituted a temporary fishing ban to allow longer-sized stocks to rebuild.

When buying fresh anchovies of any provenance, a shiny, silvery colour and a clean marine smell are good indicators of freshness. Torn fish, excessive messiness, hints of mould and, above all, any stench of rot should put you off.

Nutrition facts

ANCHOVY, EUROPEAN, RAW
per 100 grams

ENERGY (kcal)	131
PROTEIN (g)	20.4
CALCIUM (Ca) (Mg)	147
IRON (Fe) (Mg)	3.3
ZINC (Zn) (Mg)	1.7
SELENIUM (Se) (μg)	37
VITAMIN A (RETINOL) (μg)	15
VITAMIN B12 (μg)	0.6
EPA (g)	0.538
DHA (g)	0.911

I know you from pizza, right?
Probably. I end up there a lot, although I'm not sure that's the setting that I find most flattering.

Where would you rather be seen?
I don't have to be seen. If you dissolve me in a sauce, or grind me into a vinaigrette, or emulsify me in a *bagna cauda* with cream and garlic, I provide depth of flavour without anyone being able to spot me. Perhaps that's the kind of influence I relish most.

Do you get mixed up with a sardine much?
Sadly, it does happen. But we're from different families – Clupeidae for the sardine, Engraulidae for me. The sardine looks a little bloated. I'm slimmer. My flesh is darker. Also, you asked about pizza. Have you tried putting a sardine on a pizza? I didn't think so.

You sound a little vain for someone your size.
Remember that I speak for my tribe. We travel in huge packs, and there's strength in numbers. Around the Mediterranean, we've wowed the crowds for thousands of years. And we pack more flavour than sardines, though of course I wouldn't want to disparage anyone. We cost more too.

Unfortunately, no sardine is here to defend itself, so why not send our readers a constructive message instead?
Please teach your children to like me! I'm nutritionally great for kids (though do watch the salt content if I'm preserved).

THE INTERVIEW
ANCHOVY

Anchovy Pilaf

HAMSİLİ PILAV

"The Turks are the greatest Mediterranean connoisseurs of the anchovy," writes the British diplomat and cookery author Alan Davidson. The tone of his *Mediterranean Seafood*, a book both encyclopaedic and quirkily humane, may seem a little peremptory half a century after publication. Even so, there's no denying that the Black Sea anchovy – *hamsi* – is central to Türkiye's food and socializing culture. The country produces over 170 000 tonnes of anchovies a year, with the season running from November to February: fresh *hamsi* are a winter delicacy.

Anchovy fisheries are particularly dense in and around the port city of Trabzon. The area is home to Türkiye's Central Fisheries Research Institute (SUMAE) – but also, more ancestrally, to a seam of anchovy-themed lore: this includes the Horon folk dance, in which shimmying movements are said to replicate the thrashings of the captured *hamsi*.

As early as the first half of the seventeenth century, the Ottoman traveller Evliya Çelebi marvelled at Trabzon's *hamsi* obsession. Residents – many from the Georgian Laz community – prepared the fish, Evliya wrote, in some 40 different fashions. Among these, he notoriously listed a sweet anchovy baklava. If true, that intriguing recipe appears to have been lost. Then again, the travel literature of the time did not shrink from the odd flight of fancy: Evliya also tells us, for example, that all bodily pains disappear right after consuming anchovy. Our *hamsili pilav* might not quite hit that target (if it does, we'd like to hear), but should hit a few pleasure centres nonetheless.

Hamsili pilav is unambiguously a party dish; you may think of it as a sea torte. It consists of cooked rice, fragrant with currants and spice, which is then encased in anchovy fillets and cooked again.

If you buy your anchovies whole, you can fillet them by tearing off the head, then running your thumb down the cavity to flush out the innards. When this is done, pinch the top of the backbone between forefinger and thumb and pull down. The bone will come off with the tail, leaving behind naturally butterflied fillets.

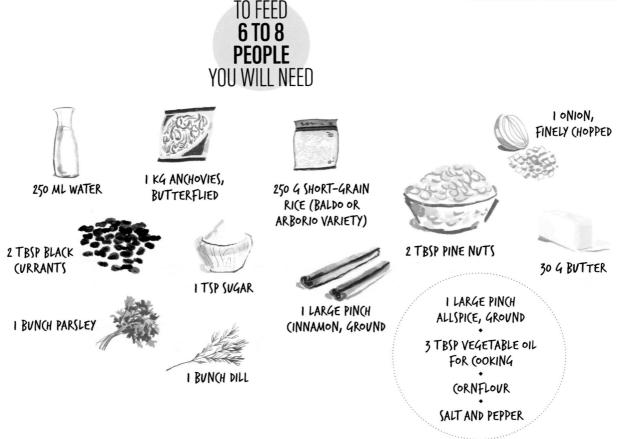

TO FEED
6 TO 8
PEOPLE
YOU WILL NEED

250 ML WATER

1 KG ANCHOVIES, BUTTERFLIED

250 G SHORT-GRAIN RICE (BALDO OR ARBORIO VARIETY)

1 ONION, FINELY CHOPPED

2 TBSP BLACK CURRANTS

1 TSP SUGAR

2 TBSP PINE NUTS

30 G BUTTER

1 BUNCH PARSLEY

1 LARGE PINCH CINNAMON, GROUND

1 BUNCH DILL

1 LARGE PINCH ALLSPICE, GROUND
·
3 TBSP VEGETABLE OIL FOR COOKING
·
CORNFLOUR
·
SALT AND PEPPER

METHOD

1 Cover the rice with water, leave to stand for half an hour, then strain.

2 Heat the oil gently in a pan, then tip in the chopped onion and pine nuts and sauté until the onion has become translucent and the pine nuts have coloured, taking care not to burn them.

3 Add the rice to the pan and sauté for another 3–4 minutes, then pour in the currants, cinnamon, ground pepper, allspice, sugar and a good sprinkling of salt.

4 Pour the water on top, give it a stir and bring to the boil, then turn the heat down and continue to cook until the water has evaporated. Take the rice off the flame, stir in a handful of chopped parsley and dill, then set aside to cool.

5 Separately, butter a medium-sized round baking dish. Sprinkle your anchovies with salt and roll them in flour; then, starting from the middle of the baking dish, begin arranging them in a star-shaped pattern, row after row, fanning out till you reach the edges of the dish. Make sure you fill in all the spaces. Continue to line the sides of the dish with anchovies, going up and leaving no space between them, until the anchovies hang off the top of the dish.

6 Spoon the rice into the anchovy-lined dish until you reach a few millimetres from the top, packing it tightly and smoothing it so it's flat. Fold the hanging flaps of the anchovies in over the rice, then continue to arrange the remaining anchovies in a star shape, working your way from the edges of the dish to the centre this time, until all the rice is covered.

7 Bake the *hamsili pilav* for 25–30 minutes until the anchovies are fried and sizzling. Rest for a few minutes, then turn it out of the baking dish. Carve it up like a cake and serve.

Carp

CYPRINIUS CARPIO

So established and ubiquitous is the carp across sections of Europe, the Near East and Asia that it is durably woven into rituals and representations. In much of central Europe, but especially in landlocked Czechia and Slovakia, carp is the centrepiece of the Christmas meal. In Poland and Romania too, which despite long stretches of coastline lean inland cuisine-wise, carp, an eminently freshwater animal, is the default fish. For Poles who go by the book, it is one of 12 prescribed Christmas courses, slotted between *barszcz*, a wine-red beetroot soup, and plump *pierogi* stuffed with cabbage and mushrooms. Most Romanians cleave to pork for their Christmas indulgence. But until diets diversified in the post-communist era, they would eat carp at any time of year, done any which way: baked, breaded and fried, in aspic or in *saramură* – a briny broth flavoured with laurel, garlic, black peppercorns and roast capsicums, to which the separately grilled fish is added before the lot is served over steaming boiled cornmeal.

There seems to be, in fact, a carp curtain dividing the world, its contours determined by both cultural and wealth factors. This generally economical fish has poor traction with western European palates, drawn to maritime more than riverine flavours: carp tastes of lakes and ponds, of the Danube, of the great waterways of Ukraine and the Eurasian landmass. In North America, carp is virtually unknown as a food fish, and frequently considered a pest. Conversely, the further east you travel, the higher carp's status and the stronger the demand. More carp is farmed in Asia – millions of tonnes – than any other fish. In China, carp is a restaurant staple, served sizzling with soy sauce, ginger and green onions, or else bubbling in a fiery Sichuan-style hotpot. The larger "bighead carp" (*Hypophthalmichthys nobilis*) is the most prized species among Chinese diners, and the animal's head, intuitively enough, the most sought-after part.

A recipe for success

As well as much eaten, the carp is much depicted. Chinese art brims with exquisite scenes of ornamental carp leaping high over the water surface, in an allegory of strength and resilience, or swimming in pairs to symbolize harmony. Regional folklore assumes a continuum between the carp and the dragon: by vaulting over a gate, the former is able to morph into the latter, connoting a valiant dash for excellence. The image was historically a proxy for success in China's civil service exams; to this day, carp emoji are shared there as a digital good luck wish.

The Amur carp (*Cyprinus rubrofuscus*) in particular has been aesthetically cherished for over a millennium in China, and bred commercially for colour mutations since around 1920 in Japan. Even so, carp's positive associations in both nations are in no small part due to phonetics. Both the Chinese and Japanese language lend themselves to punning and homophony. In Chinese, the word for "carp" (and fish in general) sounds much like the word for "abundance," *yù*. In Japanese, "carp" and "love" are rendered identically as *koi*.

Now, red koi may please the eye and silver carp jump high, but most species could hardly be seen as reaching for the sublime. The common carp is a placid bottom feeder, using its toothless mouth to suck nutrients from the floor of river, pond and lake. This roiling and muddying of the waters, technically known as turbidity, can disrupt other forms of aquatic life: it has led to vigorous attempts

to eliminate carp from river systems, such as those of the United States of America, where they are non-native. These days, in fact, most carp will come from aquaculture – and that's where, ideally, yours should come from, unless you've fished it yourself.

There's frankly a lot to like about carp. The flesh is white, firm to flaky, and one of the best sources of essential fatty acids and fat-soluble vitamins – A, E and D. Carp live long and reproduce with gusto, which makes farming them an excellent way to improve food security and incomes in one neat package. In Madagascar, one of the world's least developed countries, FAO promotes the breeding of carp in rice paddies. On top of saving farmland, this boosts the yields of the plant and provides a good habitat for the fish.

An ally for food security

Overall, though, African production is heavily concentrated on the continent's eastern Mediterranean fringes, in Egypt. In most of sub-Saharan Africa, local alternatives prevail. They include the central African *mboto* (*Distichodus antonii*), which you'll find in our recipe selection. Also known as yellow-fleshed carp, *mboto* is not technically that, but comes close in both flavour and social function: where incomes are lower and access to protein more precarious, *mboto*, like carp, is a nutritional friend in need.

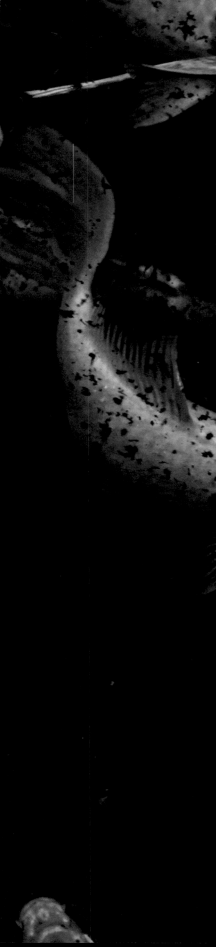

KNOW YOUR FISH

There is little fraud or mislabelling to fear when buying common carp. The fish is both inexpensive and – since it's overwhelmingly bought whole – pretty unmistakable with its round, thick-lipped mouth: it sits on the head like the puckered end of a balloon. A regular, grey-brown lattice pattern marks the body from gills to tail. (The fact that carp is seldom sold pre-cut further explains its near absence from the US market, where fish fillets dominate.)

You will most likely come across carp measuring 30–60 centimetres, which makes this a flexible option in terms of the numbers you can feed. The scales are thick: have your fish cleaned well and watch out for small bones.

Carp is sometimes described as tasting "muddy," but this is extrapolated from its bottom-dwelling behaviour. The flavour is decidedly robust. Rub the fish with lemon and remove the darker parts to lessen it. Any remaining unpleasant taste is largely down to histamines, a stress response to rising body temperatures if the carp isn't iced straight after harvesting. This brings us to the key point: carp must be achingly fresh. In central Europe, it was common until recently for the fish to be bought live and kept swimming in the family bathtub until ready to cook. You needn't go to such lengths, clearly. But if you're buying carp for dinner, best make sure the fish was still alive while you were having breakfast.

Nutrition facts

CARP, RAW
per 100 grams

ENERGY (kcal)	127
PROTEIN (g)	17.8
CALCIUM (Ca) (Mg)	41
IRON (Fe) (Mg)	1.2
ZINC (Zn) (Mg)	11.5
SELENIUM (Se) (μg)	13
VITAMIN A (RETINOL) (μg)	9
VITAMIN D3 (μg)	25
VITAMIN B12 (μg)	1.5
EPA (g)	0.238
DHA (g)	0.114

What recommends you for a career in feeding the world?
I'm highly adaptable and I mature quickly. Clearly, a little warmth helps: in subtropical and tropical areas, I can grow twice as fast as in temperate zones. But also, you can grow me in a pond with other species. Chances are, I'll do fine. And everyone will do better, in fact.

That's called polyculture, if I'm not mistaken?
Correct. The idea is that in an undrainable pond, you mix and match species from (forgive me if I get a bit technical here) different trophic and spatial niches. You have various natural fish food organisms – phytoplankton, zooplankton, detritus, plus the fish manure – at different strata of the pond water column, as well as at the bottom, where I tend to feed.

I suppose the idea of a bottom feeder may put some people off.
But that's nonsense. As I rummage in there, I aerate the pond for everyone else. And I help control the algal biomass. Trust me, everyone does well out of it. Production is higher per unit area than in monocultures. Clearly, you have to do it right, but I trust FAO with this sort of thing: they've honed it quite well and have spread that knowledge around a great many projects.

What about the notion that you taste muddy?
Inaccurate and insulting – if I'm fresh and iced properly.

Tell me something intimate. A detail.
I have a sweet tooth. In Eastern Europe, where people used to keep me swimming in tubs until it was time to... you know...

Yes?
Well, children would sometimes sprinkle a little sugar on the water surface for me. Christmas is for everyone.

THE INTERVIEW
CARP

Fisherman's soup

HALÁSZLÉ

Lying flat in the middle of Europe, Hungary is core carp country – as well as one of the world's most water-savvy nations. Lake Balaton has the status of an inland sea; the Danube and the Tisza shape the folklore and *terroir*; and hot springs bubble through Budapest's cityscape. All of this has built up to a distinct hydrological expertise – for engineering and therapeutic purposes, but also for winemaking and aquaculture. The first fish farms were established here in the 1890s. Add Hungary's status as the home of paprika, and you'll find these strands gastronomically synthesized in the hot red *halászlé*. Historically cooked by fisherfolk riverside, *halászlé* is now present on tables around the country.

The soup has its variations: the version associated with the town of Szeged, on the Tisza, involves a mix of carp, pikeperch and catfish – as opposed to the carp-only Danubian iteration. Fish roe is sometimes incorporated for extra layering and texture, as are, occasionally, egg noodles. But what *halászlé* boils down to is a rich, spicy stock made with chopped freshwater fish (skin, bones and all), tomatoes and hot peppers. The soup is then strained and augmented with the finer bits of the fish. Much like in the Sichuanese hotpot, carp's sturdy flavour stands up well to the fiery broth.

Serve the *halászlé* with chunks of white bread, good for dunking and soaking, and a glass of Riesling or white wine spritzer (*fröccs*).

TO FEED
4 PEOPLE
YOU WILL NEED

1 CARP (1—1.5 KG)

1 ONION,
FINELY
CHOPPED

1— 2 TOMATOES,
DEPENDING ON SIZE,
CHOPPED

2 TBSP PAPRIKA
(hot, sweet or smoked, or a mix,
according to taste)

1— 2 CHILLIES, DESEEDED

METHOD

1 Have the fish gutted and filleted – but keep head, tail, fins and bones as well. (Or do it yourself, but this could be messy.) Cut the fillets into 3-cm thick slices. Salt them and reserve, well refrigerated.

2 In a large saucepan, fry the chopped onion in oil till golden. Add the paprika and quickly stir to coat the onion, ensuring it doesn't burn.

3 Add the fish head, bones and other odds and ends and cover with water. Season with salt and pepper, bring to the boil and let simmer for an hour or so, until the flesh comes off the bone.

4 When the broth is ready and flavoursome, strain through a colander. Throw out the fish ends, but keep any bits of flesh that have fallen off or that you can pick from the carcass. Grind this extra flesh, manually or mechanically, and pour it back into the broth to thicken it.

5 Return the broth to the pan and bring back to the boil. Now add the fish fillets to the broth, together with the chillies and the tomatoes. Turn the heat down and simmer for another 15 minutes or so till the fillets are soft. Don't stir too much, or the fillets might collapse. Serve hot.

Fried fish with garlic and herbs

QOVURILGAN BALIQ

Uzbekistan features cooking that is frank and fragrant, a moreish crossroads of Slavic, Turkic and Asian culinary traditions. Dill meets coriander here, while lamb and raisins join parsnips and beets. Uzbek melons once made precious gifts for caliphs and czars. So popular is the nation's cuisine across the former Soviet space that Moscow alone boasts hundreds of Uzbek restaurants.

That said, fish does not loom large in Uzbek diets. The fisheries and aquaculture sector only employs a few thousand people, in a population of nearly 35 million; production and per capita consumption are low. But what there is, is worth the stop. In this dish served at roadside eateries, the carp is cut to order, deep-fried in bubbling oil (be very, very careful when cooking this at home!) and immediately smothered in garlic marinade and herbs. The clash of the hot and the wet, the sizzling and the raw, the oily and the fresh – all shot through with the tang of allium – will have you begging for seconds.

TO FEED
UP TO
4 PEOPLE
YOU WILL NEED

I CARP, I— 1.5 KG,
CLEANED, GUTTED, HEAD
AND TAIL REMOVED

UP TO IO CLOVES GARLIC,
DEPENDING ON TASTE
AND TOLERANCE, FINELY
CHOPPED

2 MEDIUM-SIZED TOMATOES

I LARGE BUNCH DILL
(and parsley and coriander if
desired), finely chopped

I SQUEEZE TOMATO PASTE
(OPTIONAL)

I CAPSICUM
•
DUSTING OF FLOUR
(optional)

METHOD

1 Slice the fish across the cavity into V-shaped pieces. (This is known as a "darne" cut.) Reserve.

2 Mix a third of the garlic with half the dill (and other herbs if using) and a pinch of salt. Roll the fish pieces in the mix and leave in the fridge for an hour.

3 Steep another third of the garlic in some warm water, adding a bit of salt and vinegar, to make the marinade. Chill in the fridge.

4 Put the tomatoes and the capsicum through a grater or grinder. Add more dill, some water and the squeeze of tomato paste if using, and stir to make the dip. Season as needed.

5 Now heat cooking oil in a deep saucepan or wok. Take the fish pieces out of the fridge, dust them with a little flour if using (you don't have to, but it will add extra crispiness) and, once the oil bubbles, fry till golden. You may want to do this in stages to avoid a drop in the oil temperature. When done, take out the fish pieces and place them briefly on kitchen paper to drain off the excess fat.

6 Almost immediately – or as soon as possible – douse the fried fish in the garlic marinade. Ideally, the pieces of carp should sizzle as the garlicky liquid hits them.

7 Sprinkle the remaining dill and any other herbs over the fish. Serve quickly with the tomato dip as a side dish.

Carp stew

MACH KOLA

Moulded by rivers and home to the mighty Ganges-Brahmaputra-Meghna delta, Bangladesh is one of the world's top producers from inland fisheries. The country has a labour force to match: over a million people work in the sector. Fish is the most widely consumed food from animal sources, and carp dominates aquaculture. The popular saying has it that *mache bhate Bangal* – fish and rice make a Bengali.

In its authentic incarnation, the recipe featured here calls for a carp species called rui or rohu (*Labeo rohita*). Native to South Asia, rui is popular from Pakistan to Viet Nam. This, however, is specifically a Chakma dish, originating among Bangladesh's Indigenous minorities. It's also a fine example of *cucina povera* that shines through vibrant seasoning, rather than through a multitude of ingredients or complexity of treatment.

If you're cooking this outside the region, substitute any type of carp for the rui – or indeed, any none-too-bland white fish. Serve the *mach kola* with (you've guessed it) rice.

TO FEED 2 PEOPLE YOU WILL NEED

4 PIECES CARP FROM THE MIDDLE OF THE FISH
about 5 cm thick, cut through the bone

1 MEDIUM ONION, CHOPPED

3 CM GINGER ROOT, PEELED AND GRATED

2 CLOVES GARLIC, GRATED

2–4 GREEN CHILLIES, HALVED AND DE-SEEDED (DEPENDING ON TASTE AND TOLERANCE)

2 TBSP OLIVE OR VEGETABLE OIL FOR COOKING

1 BUNCH FRESH CORIANDER

2 TBSP TURMERIC POWDER

METHOD

1 Blend all the vegetables with salt and turmeric and rub this mix all over the fish. Leave to stand in the fridge for an hour.

2 Pour the oil into a deep pan and add the fish and marinade, with enough water to almost cover the fish.

3 Cook on a medium flame for 10-15 minutes, until the fish has become tender and the water has reduced by two-thirds. Chop the coriander and sprinkle it over the finished dish as the rice absorbs the fragrant sauce.

Barbecued leaf-wrapped fish

LIBOKE DE POISSON FRAIS

Mboto – sometimes rendered as *mbutu* or *mboutou* – is the Lingala name for a fish of the genus Distichodus, endemic to central Africa. Its members frequently resemble the stripier species of carp; some are exported for their ornamental value. In both the Republic of the Congo and the Democratic Republic of the Congo – but also in Cameroon and the Central African Republic – *mboto* provides a vital source of protein. With access to refrigeration scarce, the fish is often smoked. One survey found that in 2019, nearly nine in ten households in the city of Brazzaville were consuming smoked *mboto* and similar river fish at least once a week.

This recipe, collected from the riverside municipality of Nsele, just east of Kinshasa, uses the fish fresh. *Liboke* (again, a Lingala word) is the method of cooking fish, and other foods too, by wrapping them in leaves. These may be banana or maranta, a plant whose root is eaten and whose sturdy, decorative green parts are commonly used as packaging. The wrapping is discarded at the end, so any other thick leaves or barbecue-proof parchment paper will do: the idea is to keep the fish moist, even as it absorbs the charred scent of the barbecue. (Yes, you will ideally need a barbecue for what is essentially an easy party dish; as for the fish, substitute either common carp, other freshwater fish or, ultimately, any fish you like.) Count 200 grams of fish per person – overall quantities will depend on the size of your party and that of your barbecue.

34

YOU WILL NEED

CARP OR OTHER
FRESHWATER FISH

GINGER

SQUEEZED LEMON JUICE

ONION

TOMATOES

CHIVES

VEGETABLE OIL
•
CHILLIES (OPTIONAL)
•
LARGE STRONG LEAVES
OR PARCHMENT PAPER
FOR GRILLING

METHOD

1 Clean or have the fish cleaned and gutted thoroughly, discarding fin and bones. Dice it all up.

2 Chop up all vegetables and herbs, together with the chilli if using – de-seeded if you like your food mildly spicy. Pour over the diced fish and mix well, adding some oil and a bit of water. Season with salt.

3 Divide the fish, together with the marinade, into portions and place each portion onto a leaf or section of parchment paper. Gather the leaf or paper around the fish, leaving it space to breathe, and spear the package closed with a toothpick or chopstick, or – if using paper – tie it up with string.

4 Place on the hot barbecue for 10 minutes or so. Check one package to make sure you're not overcooking it – the fish should keep its juiciness. Serve the liboke with cassava, known locally as *chikwang*, or mashed sweet potato.

Catfish

SILURUS GLANIS, ARIUS AFRICANUS, PANGASIUS BOCOURTI

....

Scaleless and heavily bewhiskered, catfish (order of Siluriformes) struggle to convey gracefulness.
With nearly 3 000 species spread across continents – a pattern called "cosmopolitan distribution" – generalizations are tricky. But overall, these mostly (though not always) freshwater animals are clunkily big, with none of the swagger of other large fish such as tuna. The eyes are small, almost an afterthought: the food detection function is largely devolved to the mighty whiskers, technically known as barbels. These guide the fish through the semi-opaque depths where they tend to feed. Catfish lack the sizeable air-filled bladders that promote floating: theirs are comparatively tiny. Flat, bony heads further weigh the animals down.

What with all this rummaging, a "muddy" reputation clings to river catfish – even more so than to carp, a fairly close relative. When, in late 2021, engineers seeking to unblock a drainage pipe in the Austrian city of Linz brought to light a 2.5-metre, 100-kilogram specimen, the episode appeared to synthesize the creature's least cuddly attributes: lumpy, drawn to murky habitats, a potential nuisance.

All true, no doubt, but also selective. Catfish, in fact, are probably more than averagely intelligent by fish standards. We may not know what they say to each other, but we do know that their communication system, both acoustic and olfactive, is quite developed: they are good at emitting signals of distress, for example, and good also – unlike some of us lot – at gauging the precise age, reproductive state and social status of prospective partners.

Catfish are cunning, innovative predators. In 2012, one study of Europe's largest freshwater fish, the wels catfish (*Silurus glanis*), documented what appeared to be new "beaching" tactics by the animals in the river Tarn, in the French city of Albi. Lying in wait at the edge of the water, the catfish would leap onto land, in flash incursions, to snatch live pigeons. Such hunting ingenuity, akin to the amphibious dexterity of crocodiles, testifies to highly adaptive qualities. Warm water environments, particularly those that have been modified by humans, are especially congenial to catfish: they live long lives, up to 80 years, and reproduce abundantly – all of which means catfish are easily farmed, generally sustainable to eat, and pretty inexpensive. Consequently, they are common in the diets of Europe, Africa, Asia and the southern United States of America.

Bulky but clever

When it comes to cooking freshwater catfish, you can readily eliminate any muddiness by cutting out and discarding the darker central parts of the raw fillets, then soaking the fillet strips in lemony water. As you would with carp and other bottom-feeding fish, make sure all the blood has drained off. You'll end up with flesh that is mild-flavoured and moist, loaded with lean protein and rich in vitamins D and B12: both are vital nutrients *and* frequently missing from diets.

KNOW
YOUR FISH

With catfish cheap and easily available, the risk of fraud or mislabelling at the market stall or on your restaurant plate is minimal. If anything, it is certain varieties of catfish that are more likely to be passed off as higher-end oceanic fish, such as cod or haddock. (The flesh is similarly whitish to light pink in colour, and equally flaky – if bonier for some cuts.) Freshness is paramount: like other freshwater fish, catfish spoils fast and must be kept on ice until the last moment. Sour odours, sliminess or an excessively reddish tinge are signs for you to run or call the authorities. Smoked catfish, as featured here in recipes from Guinea and Nigeria, will normally be sold cured and flattened, and may need rehydrating.

Nutrition facts

PHILIPPINE CATFISH, WHOLE, RAW
per 100 grams

ENERGY (kcal)	106
PROTEIN (g)	16.8
CALCIUM (Ca) (Mg)	58.7
IRON (Fe) (Mg)	0.8
ZINC (Zn) (Mg)	0.7
IODINE (I) (μg)	22
SELENIUM (Se) (μg)	32
VITAMIN A (RETINOL) (μg)	44
VITAMIN D3 (μg)	[1]
VITAMIN B12 (μg)	4.8
OMEGA-3 PUFAS (g)	0.35
EPA (g)	0.08
DHA (g)	0.06

[ ] = lower quality data

Are you some sort of carp?
No, but I can see why you might think that. We're both voluminous, bottom-dwelling freshwater fish, although we have some marine or brackish-water incarnations. The carp has scales. I have none; I have barbels.

And it really is you, is it? The English word "catfish" also refers to someone who impersonates others online.
I don't go online. You shouldn't make assumptions just from the fact that I'm speaking to you – although I *am* told I'm fairly clever. Humans have now figured out that we fish retain information for years, and that some of us catfish, in particular, can remember human voices and the colour of things for very long periods.

Yes – and I believe that sensorially too, you're quite advanced?
I certainly have a finer sense of taste than you do. A human individual has around 10 000 taste buds. We have several times that – the larger ones among us may have 15 times as many.

I'm suddenly beginning to wonder if the roles shouldn't be reversed. In other words, if you shouldn't be eating us.
It's not for me to say. But yes, we live in one of many possible worlds. I do find myself pondering what life might have been if the tables were turned. Still, you should watch yourself. In the Mekong river, you've nearly wiped me out. Being at the top of the food chain doesn't give you a free pass, you know...

THE INTERVIEW
CATFISH

Afghani fish

TARZ TAHIA MAHI AFGHANI, طرز تهیه ماهی افغانی
DA AFGHANI MAHI PAKHLI TARKIB, افغانی د ماهی پخلی ترکیب

Afghanistan's legacy of upheaval and the civil collapse of 2021 have combined with flooding to cause tremendous food insecurity. At the time of writing, FAO estimated that nearly half the country's people – 20 million or so – were facing acute hunger. Four-fifths of those most at risk are in the countryside: the better-resourced cities have maintained a degree of economic life, which includes lively fish markets (*machli bazaar*). Here, as scalers speedily clean the fish, much of it imported from Pakistani farms, cooks chop it up and throw it into bubbling cauldron pots.

Our own recipe features a more affluent and urbane version of Afghan fish fare: it requires some dedication in securing the ingredients, though subcontinental grocery shops will routinely store most. The charcoal will be sold in barbecue bags: in many parts of the world, these may not be available through the year. If that's the case, save this dish for a summer's day – you wouldn't want to miss out on its beguiling smokiness.

TO FEED 4 PEOPLE YOU WILL NEED

1 TBSP POWDERED CORIANDER

FISH FILLETS, EITHER CATFISH OR ANOTHER WHITE FISH
cut into 5-cm pieces

10 CASHEW NUTS

ML CANNED PUREED TOMATOES

2 MEDIUM ONIONS, QUARTERED

8 GARLIC CLOVES

1 TBSP FRESH CREAM

2 ½ TSP GHEE (CLARIFIED BUTTER)

1 TBSP YOGHURT

1 PIECE FRESH GINGER,
about 2.5 cm long, sliced

SOME FRESH FENUGREEK LEAVES OR ½ TSP DRIED FENUGREEK
•
MORE FRESH GINGER AND 1 SMALL BUNCH CORIANDER FOR GARNISHING
•
4 PIECES BARBECUE CHARCOAL

2 TSP RED CHILLI POWDER

1 SPLASH VINEGAR

4 CLOVES

1 GOOD SQUEEZE LEMON JUICE

METHOD

1 Place the fish pieces in a bowl. Add salt and lemon juice. Leave to marinate for 10–15 minutes.

2 Whizz the onion, garlic, ginger and cashew nuts in a blender, adding enough water to make a paste.

3 Separately, warm the ghee in a wok or non-stick pan. Scoop in the mixed paste and sauté for 3–4 minutes or until fragrant.

4 Add the pureed tomatoes and cook down, over a medium flame, for 8–10 minutes. Add in the chilli powder and powdered coriander. Season with salt and bring the mixture to a boil.

5 Now put in the fish pieces and, depending on their thickness, cook for another 5 minutes or a little longer. Throw in the fenugreek leaves or dried fenugreek and fresh cream. Mix well, then turn off the heat.

6 Separately, put the pieces of charcoal in a steel bowl, then place the bowl inside the wok or pan, making sure not to mix it with the food. Add the cloves over the coal. In a small milk pan, warm up the remaining ghee, then pour that over the cloves and coal. Cover the entire dish and allow the smoke to infuse it for 3–4 minutes.

7 Remove the coal bowl from the wok. Your Afghan fish is ready to serve, garnished with extra ginger and fresh coriander.

Smoked catfish casserole

POISSON CHAT FUMÉ EN SAUCE / KONKOÉ TOURÉ GBÉL

Smoked catfish is much enjoyed in Guinea. The country's late president, Lansana Conté, is reputed to have always travelled with a handy supply of it. Guinean expatriates are used to receiving shipments of it from family back home. Unlike most catfish seen globally, the species used here is a marine one, *Arius africanus*, known in French as *mâchoiron* – or "big-jawed one" – and in the local Susu language as *konkoé*. *Touré gbél* designates the sauce, liquid enough to be halfway to a soup, which accompanies the fish: its colour and warmth are derived from the red palm oil that is also present in Afro-Brazilian cuisine. (For a Brazilian recipe involving the local version of palm oil, see p. 157.)

This dish would be served quite spicy in Guinea: dial the chilli content down – or up, for that matter – to suit your taste. The recipe calls for manioc (see cassava roots on p. 160), but if this is hard to come by or too arduous to prepare, replace it with sweet potatoes or yams.

TO FEED
4 PEOPLE
YOU WILL NEED

4 CARROTS, SLICED
MEDIUM

I KONKOÉ
(SMOKED CATFISH)

or other smoked fish
of about 600 g

250 ML RED PALM (OR
VEGETABLE) OIL

2 POTATOES,
DICED SMALL

3 TOMATOES,
FINELY CHOPPED

I CLOVE GARLIC,
FINELY CHOPPED

2 MEDIUM-SIZED
AUBERGINES, PEELED

2 YAMS,
DICED MEDIUM

2 ONIONS,
FINELY CHOPPED

I SMALL BUNDLE CHIVES,
FINELY CHOPPED
•
SALT, PEPPER
AND CHILLI TO TASTE

5 BASIL LEAVES

METHOD

1 Wash the fish well in warm water to soften it, then flake it into bite-sized pieces (or a little larger). Heat up the oil in a saucepan, add the aubergines, skin on, and sear them on all sides. Turn the flame down a little and continue to cook the aubergines.

2 Grind the onion, garlic, tomatoes and herbs in a mortar or mixer, with salt, pepper and chillies if using. Take the aubergines from the pan when they're soft, making sure you don't burn yourself, add them to the mortar or mixer, and blend them in.

3 Return the ground mix, aubergines included, to the hot oil. Add enough water to obtain a loose sauce and bring to a low boil. Then add the flaked fish to the pan, together with the rest of the vegetables. Bring back to the boil and simmer for 10 minutes or so, or until the potatoes are done.

4 Adjust the *konkoé* for seasoning and serve over white rice.

Inle fish steamed in banana leaves

AAINNLAYY NGARRHTAMAINNNAA,
အင်းလေးငါးထမင်းနယ်

Southeast Asia is home to the Mekong giant catfish, a critically endangered species in the wild. This 3-metre-long fish, to quote one *American Scientist* piece from 2004, "grows as fast as a bull and looks a bit like a refrigerator". While official protection has been insufficient to stamp out poaching, the last few years have seen stepped-up efforts, particularly in Viet Nam, to limit illegal, unreported and unregulated fishing. Viet Nam is one of the world's aquaculture powerhouses: it supplies ninety percent of all catfish imports into the United States of America.

Farmed basa (*Pangasius bocourti*), commonly sold as "pangasius," is consumed in myriad ways across Southeast Asia. So is the "walking" or Philippine catfish (*Clarias batrachus*). In Thailand, the famed cookery teacher and writer Srisamorn Kongpun makes spicy stir-fried catfish (*pad prik khing pla-dook foo*) by crumbling the fish and sprinkling it into

hot oil, then pressing it together in the pan to form a sheet, and further cooking this sheet on both sides: the method ensures that the fish is crispy within and without. Our own recipe comes

from neighbouring Myanmar. In it, the catfish – sourced from Lake Inle, rich in endemic species – is steamed and smothered in fragrant lemongrass, coriander and ginger.

TO FEED
2 PEOPLE
YOU WILL NEED

10 MEDIUM SPRING ONIONS

330 G CATFISH FILLETS
(or other white fish)

4 CLOVES GARLIC

2 STICKS LEMONGRASS

5 CM FRESH GINGER

I HANDFUL CORIANDER

1 TBSP RICE (UNCOOKED)
•
GENEROUS PINCH SALT

METHOD

1 Toast the rice lightly in a dry pan, making sure not to burn it, then grind it into a coarse powder.

2 Cut the fish fillets into strips. Separately, crush the garlic with the ginger, chop the spring onions, coriander and lemongrass (discard the outer skin of the lemongrass if it's too tough), and add in the toasted rice, mixing well.

3 Toss the pieces of fish through the mix. Divide this fish mix into two portions and place each portion onto a banana leaf or section of parchment. Bring up the corners of each leaf or section of parchment to form a parcel, then spear through the top with a skewer to close the parcel.

4 Steam the fish parcels over boiling water for 15 minutes or so. Serve the contents over rice or on its own, as a warm salad.

Catfish egusi soup

Unless you get to cook this umami-rich recipe in its homeland, you may need to visit a specialist grocery shop or root around online for much of what goes into the pot. Some of the ingredients – the smoked fish, red palm oil and fermented locust beans – sit on the western African culinary continuum; others are more specifically Nigerian. The beans, known as *dadawa* (or *iru* in Yoruba), are a punchy, fermented taste enhancer analogous to miso, Marmite or Vegemite, yet also evocative of shrimp paste. Ground crayfish, another local condiment, further brings out these fishy notes. The fluted pumpkin leaves (*ugu*) bristle with vitamin C and iron, on top of their pleasingly green tinge and bitter sapidity. The eponymous egusi, finally, refers to a nutty-flavoured melon or gourd seed, used as a protein-rich thickener.

Egusi soup is traditionally eaten with "swallow" – Nigerian dough balls made with either yam or cassava flour – but rice is an established alternative. Or you may sacrifice geographic accuracy and enjoy your egusi with chunks of crusty baguette.

TO FEED
4 PEOPLE
GENEROUSLY
YOU WILL NEED

200 ML
RED PALM OIL

300 G SMOKED CATFISH

1 STOCK CUBE

1 HANDFUL UGU
(FLUTED PUMPKIN
LEAF), CHOPPED

1 TBSP DADAWA
(LOCUST BEANS)

300 G GROUND EGUSI
(MELON SEEDS)

1 CAPSICUM
•
SALT AND PEPPER
TO TASTE

1 HABANERO
or other chilli pepper, deseeded
(use only part or omit if desired)

2 MEDIUM-SIZED
ONIONS, CHOPPED

METHOD

1 In a mixer, blend the capsicum and habanero with half a chopped onion and a little water to loosen the mix. Reserve.

2 Separately, blend another half of chopped onion with the egusi and the crayfish. Set aside.

3 In a saucepan, heat the palm oil and gently fry the remaining chopped onion till translucent. Add the *dadawa* and simmer for another minute, then pour in the pepper and onion mix and cook until the mixture forms a loose paste.

4 Add the egusi, crayfish and onion mix to the pan. Thin out the mix with a little water, throwing in the stock cube and making sure it dissolves.

5 Finally, add in the chopped *ugu* leaves and the smoked catfish. Cook for 15–20 minutes, seasoning to taste and adding more water if needed. Your egusi soup is ready to serve when it's achieved the consistency of a liquid stew or dense broth.

Fried curried catfish

Fish consumption is still low in Zimbabwe. Livestock forms the basis of subsistence farming here; commercial agriculture is dominated by land crops such as tobacco. But the potential is great: the waters of the Zambezi River system, which include Lake Kariba, the world's largest man-made reservoir, are fish-rich and lend themselves well to aquaculture. FAO's efforts are mostly aimed at upgrading Zimbabwe's tilapia sector, seen as the most promising in nutritional and commercial terms. Meanwhile, local species of catfish – particularly the vundu (*Heterobranchus longifilis*) and sharptooth (*Clarias gariepinus*) – are generously sized, meaty and popular. A farmed hybrid of these, the *Hetero-clarias*, is also gaining ground. As elsewhere, use any catfish you like for this no-fuss recipe: basa/pangasius would be easy to source. Take the quantities listed here as a suggestion: this is tasty but largely empirical culinary territory, with calibration anything but prescriptive.

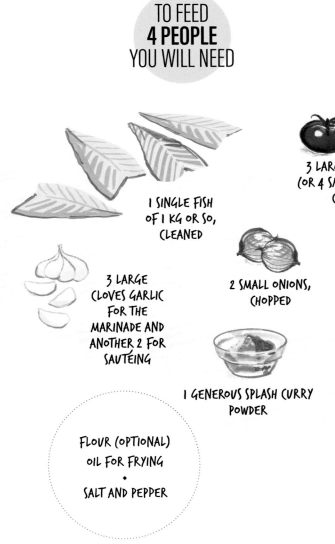

TO FEED
4 PEOPLE
YOU WILL NEED

1 SINGLE FISH
OF 1 KG OR SO,
CLEANED

3 LARGE TOMATOES
(OR 4 SMALLER ONES),
CHOPPED

3 LARGE
CLOVES GARLIC
FOR THE
MARINADE AND
ANOTHER 2 FOR
SAUTÉING

2 SMALL ONIONS,
CHOPPED

1 GENEROUS SPLASH CURRY
POWDER

FLOUR (OPTIONAL)
OIL FOR FRYING
•
SALT AND PEPPER

METHOD

1 Crush 3 cloves of garlic, mix with salt and pepper, and rub the mix all over and inside the fish. Fry the fish in a pan until brown on both sides (sprinkle it with flour previously for a crustier feel, if desired) and set aside.

2 Add more oil to the pan if needed, scrubbing in the sticky bits of fish with a spatula, and sauté the onion and remaining 2 cloves of garlic until tender, then add the chopped tomatoes and curry powder. Cook the mixture until soft. Taste it: you're looking for a sweet-spicy roast aroma. Adjust the seasoning as needed. If the tomatoes are too acidic, sprinkle a little sugar into the sauce.

3 Place the fried fish in the sauce and baste gently on a low flame until it's warmed through. Serve over rice or potatoes, and perhaps some raw or wilted spinach for greenery relief.

Cod

GADUS MORHUA

Cod's flesh, all flakes and pale as first snow, reflects its lack of muscularity. This yielding creature is the polar opposite of the purple fighting machine that is the bluefin tuna, another modern favourite. True, cod is tough in some ways, long-lived at 20 years or so, resistant to disease and remarkably thick-skinned. But it goes limp when hooked, a trait that hasn't historically done it any favours. A fish of frigid waters, it's also attracted to (slightly) warmer oceanic shores when it's time to spawn, and so practically offers itself for capture. Then again, it's not by centuries of hauling cod into artisanal fishing boats that we came close, in some areas, to wiping it out: it was by industrially scraping the seas in a matter of decades, with our greed goggles on and our conservation compass off.

"Cod" is almost invariably taken to mean Atlantic cod (*Gadus morhua*), though a Pacific cousin exists. A number of other commercial species – hake and haddock, ling and pollock – count as family members (Gadidae). Some are increasingly seen as alternatives to cod, without quite matching its prestige.

When the American journalist Mark Kurlansky published *Cod: A biography of the fish that changed the world* in 1997, it looked as though its subject might be about to exit history. The book portrayed cod as a hinge factor in the European settlement of North America. First fished by the Basques and the Vikings 1 000 years ago, then by the Portuguese, the French and the English, cod provided both food security and handsome profits. It spurred exploration and the opening of trade routes to Newfoundland and New England; was traded for slaves in western Africa and, in turn, fed to slaves toiling on Caribbean sugar plantations; drove demand for salt, needed to preserve the fish in pre-refrigeration days; caused trade wars from the seventeenth century to the twentieth; and was ultimately, the book implies, indissociable from circumstances that brought about the US Declaration of Independence.

Back from the brink

As with other commodity monographs – tea, tulips and suchlike – the singular focus of *Cod* lays it open to charges of hyperbole. But the book's documentary value is great, as is the elegiac tone of its reportage. Its merit is, not least, exogenous: a bestseller, it rippled into a wave of popular awareness, at a time when the finite nature of the planet's resources had yet to gel as a household concern. By the turn of the millennium, cod, with its storied past, appeared to be succumbing to the age of the super-trawler, and consumers knew it.

Politically, in fact, moves were already afoot to prevent cod's feared extinction. In the 1980s, both Iceland and Norway had enacted drastic fishing quotas. (Skirmishes between Iceland and the United Kingdom over access to cod-rich waters had formed a political mood music for decades.) By the early 1990s, Canada's inshore and offshore cod populations had collapsed, prompting a federal moratorium and the largest industrial closure in the country's history. As the decade ended, Canadian cod was classified as "of special concern".

One generation later, stocks have rebounded in the northeast Atlantic – but hardly (or very modestly) in Newfoundland, around its Grand Banks extension, in Nova Scotia or New England. In these areas, climate change seems to have compounded the legacy of overfishing, with a likely irreversible alteration of the food chain. Where cod were once top predators, their former prey, mainly crabs and other crustaceans, have proliferated and now fill that role. This does mean that those waters are economically viable once more: lobster, having migrated north, has probably never been as abundant as it is now off the US state of Maine. But the northwest Atlantic's biodiversity has taken a battering. Maritime Canada has seen a whole fishing culture extinguished.

Today, Atlantic cod is at its most sustainable in the freezing waters of the Barents Sea. Here, Norway and the Russian Federation have been managing stocks through a joint fishery commission, setting quotas and ensuring more or less stable population levels. In the early 1990s, data exchange and compliance inspections were stepped up. At the time of writing, that said, the war in Ukraine was straining relations between Moscow and Western allies: there were signs that sanctions were causing shifts in the global seafood trade. Political cooperation in the Arctic was breaking down, leaving the status of joint Norwegian-Russian cod quotas unclear. In the last year before the war, the quotas had been cut by a fifth.

Better managed, still wild

All of this means that in G7 countries in particular, the availability of Russian cod may dry up. (The Russian Federation also produces Pacific cod, which it manages separately.) Sanctions aside and from a sheer environmental perspective, Norwegian or Russian cod, if reliably labelled as such, can be eaten without qualms. So can Icelandic cod, following decades of sound management.

Efforts to farm cod, once powered by extinction concerns, have proceeded fitfully. There have been minor successes, especially in Norway. But with the exception of small projects, the rebound in wild stocks has largely quashed the impetus. Commercially, cultured cod remains a marginal proposition.

KNOW
YOUR FISH

Back when "zero waste" was an economic necessity for most (rather than something rich-country citizens need cajoling into for climate's sake), every last bit of the cod was used. The flesh is extraordinarily lean and loaded with protein. Most consumers will only come across fillets these days, but the throat, known as "tongue," was long thought a delicacy and still is in some quarters. Some of us may remember being given cod liver oil as children. In Iceland, Spain and Portugal, the liver can be seen tinned, and sometimes smoked, as can the roe. Cod bones found employment, in centuries past, as a natural fertilizer. The strong skin was processed into leather – but it's also delicious to eat, acquiring a wonderfully glutinous texture in the frying pan. (Remember to score the skin when cooking fillets, or it will cause the flesh to curl as it loses collagen and shrinks.)

How you find cod will likely depend on where you buy it. Northern European and "Anglo-Saxon" markets are partial to fresh cod, traditionally the stuff of English fish and chips (though hake, haddock and even sea bass have become popular alternatives). The Mediterranean and Caribbean markets prefer salt cod. A third option, called stockfish, is most closely associated with Norway: this is air-cured cod, without the addition of salt. Note that salt cod (*baccalà*, *bacalao* or *bacalhau*) must be soaked at length to rehydrate and release most of that sodium: a lot will be retained, so there's no need to salt further. Rinse the fish well and use tweezers to remove all pin bones.

Whichever way you get your cod, it will take magnificently to dairy, fats animal or vegetal, and Mediterranean herbs. In the Veneto region of Italy, *baccalà mantecato* sees the rehydrated fish simmered in milk with black pepper and bay leaves, then mashed and *montato* – thickened and made unctuous – through the gradual addition of olive oil. (In a nod to the commercial crosscurrents that put cod on European tables for centuries, this recipe, usually served with polenta, traditionally uses Norwegian-style stockfish.)

Over in French Provence, *brandade de morue* involves similarly warming the reconstituted cod over a low flame with cream, garlic, thyme and cloves, then enriching it with olive oil and beating it with floury boiled potato. Such dishes are homemade ointments for your insides: spread them thick and let them soothe you.

Nutrition facts

ATLANTIC COD, WILD, FILLET, RAW (NORTHEAST ATLANTIC)
per 100 grams

ENERGY (kcal)	80
PROTEIN (g)	18.5
CALCIUM (Ca) (Mg)	9
IRON (Fe) (Mg)	0.2
ZINC (Zn) (Mg)	0.4
IODINE (I) (µg)	260
SELENIUM (Se) (µg)	29
VITAMIN A (RETINOL) (µg)	2
VITAMIN D3 (µg)	2
VITAMIN B12 (µg)	1.1
OMEGA-3 PUFAS (g)	0.22
EPA (g)	0.06
DHA (g)	0.16

Hello. I've seen you credited with changing the course of Western history.

Not that I'd want to turn down the honour, but that's a selective reading of events. Let's put it this way: I've played a part in northern and western Europeans' urge to roam the north Atlantic, and then in the economic set-up of what were originally settler communities. So perhaps yes, in some small way I did have a hand – or fin, rather – in shaping the social profile of parts of North America.

You're being rather diplomatic. Also, you don't display anything like the ego of smaller fish I've spoken to. Do you think this self-effacement has made you a soft touch?

It's true that I move slowly and tend to go limp when caught. On the other hand, I'm carnivorous, and even cannibalistic on occasion: I sometimes eat codlings. Also, as a species, I have no natural predator, which is probably why I've never really developed a strong survival instinct. Actually, that's not quite true. I do have one predator.

I know, I know... We thought you'd last forever. We were wrong. We know better now. I hope you've noticed.

Yes, things have improved over the last couple of decades. I'd be churlish not to admit it. You finally did see the light. I don't much see it myself, as it's pretty dark deep down, where I live. And at those latitudes in winter, even the sky is a sea of gloom. There's not much to do except, ahem, the obvious: no wonder that come spring, I lay up to 500 million eggs.

THE INTERVIEW
COD

Christmas cod

JULETORSK

Norway is an A-list fisheries and aquaculture power. In 2021, it exported nearly USD 14 billion of fish and seafood products – a second place in aggregate terms, but a towering world record at more than USD 2 500 per capita. On one level, this correlates with the country's general wealth, economic performance and sound environmental stewardship. But it also reflects historic choices shaped by physical and human geography. Norway has one of the world's longest coastlines; a craggy, mountainous terrain, with little space for agriculture; and a sparsely distributed people to whom navigation was, for centuries on end, inseparable from life.

Norwegians have therefore fished cod for countless generations – since long before they even began to think of themselves as Norwegians. In northern Norway in particular, cod is a staple on Christmas eve – for those who'll have fish rather than meat, that is, as the two traditions coexist. Our recipe is (kindly and consensually) pilfered from the Norwegian Seafood Council, a public body under the Ministry of Trade, Industry and Fisheries, headquartered in the northern city of Tromsø: it blends a certain Arctic trimness with Norway's culture of festive domesticity.

Married with the spices baked into the cod, the sauce bathes the dish in memories of mulled wine: go for a light-bodied red. You may choose to ignore the sauce if you're not a drinker – though bear in mind that almost all alcohol will have been boiled out of it.

TO FEED 4 PEOPLE YOU WILL NEED

FOR THE VEGETABLES

300 G SWEDE (RUTABAGA)
•
150 G CARROT
•
I LEMON QUARTER
•
50 ML SINGLE CREAM
•
25 G BUTTER
•
I TWIG ROSEMARY

4 FRESH COD FILLETS
(around 150 g each), skin on

I PIECE STAR ANISE

2 CLOVES GARLIC, HALVED

I BAY LEAF

I ½ TSP GRATED ORANGE PEEL

I CLOVE

3 GRAINS BLACK PEPPERCORN, CRUSHED

I ½ TSP GRATED GINGER

I TBSP SALT

FOR THE WINE SAUCE

50 CL RED WINE
100 ML FISH STOCK
(If you have fish heads or other parts of the fish, make your own; it not, use a cube.)
•
I TBSP BUTTER
•
½ SHALLOT
finely chopped
•
I TSP CHIVES,
finely chopped
•
I TSP THYME,
finely chopped
•
SALT AND PEPPER

METHOD

1 Preheat the oven to 150° C. In the meantime, score the skin of the cod fillets, sprinkle salt all over them, and leave them to rest for 10 minutes.

2 In an oven-proof dish, pour the olive oil and add the spices and bay leaf, the halved cloves of garlic, the crushed peppercorns, the ginger and the orange peel, disposing them evenly along the bottom of the dish.

3 Rinse the cod fillets, pat them dry and place them in the oven dish over the spiced oil. Bake for 15 minutes or so. When they're done – they should be moist inside – gently strip back the skin. (You can save the skin and crisp it up in the oven for a snack or discard it.) Keep the fillets warm, either in the switched-off oven with the door open or in a dish covered with cling film, making sure they don't dry out.

4 While the cod fillets are in the oven, boil the carrot and swede until tender, with the rosemary and lemon.

Strain the vegetables well. Puree the carrot and swede, separately, and keep warm. Discard the rosemary and lemon.

5 Now (or while the vegetables are boiling if you've managed to synchronize everything) make the sauce. In a small saucepan, soften the chopped shallots with butter, taking care not to burn them. When they're soft and melting, add the wine and fish stock, and reduce by half over a low-to-medium flame.

6 Strain the sauce, then pour it back into the saucepan. Add the butter and whisk it in to thicken the sauce and make it glossy.

7 Time to serve. For a modern brasserie touch, cut a disc of swede mash for each diner and place it in a deep plate. Then cut a disc of carrot mash (this should be a bit thinner) and place it on top, so that you end up with a double-layered mash. Pour some wine sauce around the base of the mash. Finally, place the cod fillet on top of the mash.

Green fig and saltfish

Saint Lucia sits like a teardrop along the Lesser Antilles chain. The residents of this lush island are overwhelmingly descendants of slaves – officially English-speaking, but with French Creole elements flowing through the culture and speech. Colonial Saint Lucia changed hands for the last time in the early years of the nineteenth century. Having reclaimed the island from France, the British reinstated slavery, abolished less than a decade before in the wake of the French Revolution. Boilers that produced molasses still dot the Saint Lucian landscape, rusty mementoes of the brutal sugar plantation days.

With the cod trade booming in the 1700s, here as elsewhere in the region, salt cod became the default food fed to slaves. Mark Kurlansky notes in his book that salt, by drawing the moisture out of the fish, reduces its weight by four-fifths: this slashed transportation costs. And although

Caribbean islands only absorbed the cheapest cuts and cures, cod's high protein content kept the slaves working the fields from first to last light.

Slavery did eventually end in the 1830s. But the tradition of eating salt cod – known here as saltfish – endured. "Fig" may be misleading to non-Caribbean audiences: it refers to banana, Saint Lucia's main crop and export. This recipe is the island's national dish.

TO FEED **4 PEOPLE** YOU WILL NEED

4 CLOVES GARLIC, ROUGHLY CHOPPED

600–800 G SALT COD

3–4 GREEN BANANAS, SLICED

½ BUNCH CHIVES

1 ONION, FINELY SLICED

½ BUNCH PARSLEY

2–3 PICKLED PEPPERS, SLICED

VEGETABLE OIL FOR FRYING
•
SALT AND PEPPER

1 TBSP THYME

METHOD

1 Boil the saltfish for a good 20 minutes to rid it of the salt and rehydrate it. Taste and repeat if necessary. (Alternatively, you may soak the fish in cold water for at least a day in the fridge.)

2 Drain the fish, pat it dry and cool it. Remove the bones, using tweezers if needed, then use a fork to shred the flesh, discarding the skin.

3 Trim the top and bottom of the bananas, then cut lengthwise into the skin. Place the bananas in a deep pan, cover with water, add a teaspoon of salt and bring to the boil. Continue to boil the bananas for 15 minutes or so, until the skin has darkened and they're soft inside. When they're done and no longer too hot, peel them and keep them warm. (You may place the bananas back in the cooking water to ensure they stay soft.)

4 In a deep frying pan or a wok, warm up some vegetable oil and fry the onions, garlic and peppers over a medium flame. Now add the shredded fish and the herbs, and fry further. Season to taste. The fish should be warmed through and even a little crispy, but the dish should remain richly moist.

5 To serve, slice the boiled bananas medium thick, arrange the slices on a serving plate, and spoon the fish mix over them.

Bakkeljauw

Like its northern island neighbours, Suriname spent more than two centuries as a slave colony – of the Netherlands in this instance, which has bequeathed the modern nation its official language. In Suriname too, salt cod was a slave's staple and remains a culinary trope. But in this country where Afro-Caribbean influences meld with those of South America's native communities – not to mention more recent Indian, Chinese and Near Eastern elements – cassava (yuca) joins the fish on the plate. Alternatively, you may find the fish served in a *broodje*, a Dutch-style sandwich. The word *bakkeljauw* itself attests to Suriname's fusion of cultural and linguistic inputs: it is, of course, a Dutch rendition of the Portuguese *bacalhau*. As in our (not dissimilar) Saint Lucian recipe – and unlike in Mediterranean countries, where the salt cod is usually soaked for two or three days in cold water before cooking – the Surinamese way is to boil the salt out of the dried fish, and then fry it.

To make the cassava chips, cut into your tubers lengthwise and peel off the bark, then chop them into chunky chip shapes. Rinse them well, bring to the boil a saucepan of salted water, and boil the cassava fries in it till soft, but not disintegrating. Throw out the water, brush the cassava chunks with olive oil, and roast them in the oven at 220° C until golden, turning them over halfway through. (Reduce the temperature a little if you have a fan-assisted oven.)

TO FEED
2 PEOPLE
YOU WILL NEED

300–400 G SALT COD

1 LARGE OR 2 SMALL
ONIONS, FINELY
CHOPPED

2 CELERY STALKS

2 CLOVES
GARLIC, MINCED

2 TBSP OLIVE OIL

1 CHILLI PEPPER,
DESEEDED
(or not, depending on taste)
and minced

1 TBSP TOMATO PUREE

JUICE OF 1 LIME

1 LARGE TOMATO, DICED

SALT AND PEPPER

METHOD

1   Boil the cod in (unsalted) water until it sheds the salt and the water runs clear. You may need to change the water and start again, depending on how intensely cured the fish is.

2   Drain the fish and discard the water. When the fish is cool enough to handle, remove skin and bones, then shred it into flakes.

3   Heat the olive oil in a pan and sweat the onions in it, then add garlic and chilli, then the shredded fish. Taste and adjust for salt if needed. Stir-fry for a few minutes, then add the celery, diced tomatoes and tomato puree. Stir again.

4   Serve when the *bakkeljauw* is warmed through, with a squeeze of lime and a grind of fresh pepper. Plate it up with the cassava fries, with more chilli or a pepper relish on the side if desired, and maybe some chopped boiled egg.

Eel

ANGUILLA ANGUILLA

Freshwater eels have traversed the centuries as riddles of nature. They clogged up rivers and estuaries throughout the ancient, medieval and early modern eras. Still, no one knew where they came from or understood much about them. One myth had them self-generating from mud. Even as they gobbled them in quantities – jellying them in what is now UK territory, or braising them with a butter-and-wine *sauce matelote* in France – Europeans remained baffled by eels' slithery undulations, their general whence and whither, and their seemingly elastic lifespans. Before he turned his attention to the human psyche, a young Sigmund Freud spent weeks looking – in vain, as it turned out – for eel testicles. He dissected hundreds of the animals in the port of Trieste: how eels reproduced, in the absence of observable eggs or sexual organs, was a compulsive mind-twister.

We now know – or believe with a high degree of confidence – that European eels (*Anguilla anguilla*) come to life in the Sargasso Sea, the strangely self-contained expanse of water at the intersection of the Atlantic and the Caribbean. Their bodies transform throughout their lives as, over many years, they migrate to continental rivers and lakes, first as larvae, then as translucent "glass eels" (or elvers), then snake-like and opaque in appearance, before they return to the Sargasso Sea, their stomachs dissolving in the process, their sexual organs belatedly blooming, to spawn and, this final feat accomplished, to die.

The American eel (*Anguilla rostrata*) also breeds in the Sargasso Sea; other species, including the Japanese eel (*Anguilla japonica*), have been shown to breed at a small number of locations in the Pacific and Indian oceans – off the Northern Mariana Islands, Tonga or Madagascar. But aspects of eel biography continue to

Mystery solved?

elude us. In his learned romp through marine facts and trivia, *Eloquence of the Sardine*, the physicist and sea enthusiast Bill François recounts the story of Åle, an eel dropped in a well in the Swedish village of Brantevik in 1859. (The practice was apparently not uncommon: eels ate insects and kept the household water parasite-free.) The animal lived until 2014, its longevity likely connected to its inability to reverse-migrate and spawn – dying, it is thought, as the water in the well overheated. Åle, in other words, was accidentally cooked to death at the age of 155. Had he not been, he might yet have outlived us – and, who knows, subsequent generations too.

Europe is littered with toponyms referencing the eel, from the fishing port of Ålesund in Norway to the Italian lakeside town of Anguillara. One particularly erudite FAO staff member, who's been studying the connection between fish and heraldry, points out that Anguillara was the fief of the Orsini family, to whom a number of popes belonged: their coat of arms, still visible on buildings around Rome, prominently features the eel. Our colleague also recalls the fate visited on one pope in Dante's *Divina Commedia*: the Holy Father winds up in Purgatory because of his gluttony for eels and Tuscan wine.

While overindulgence in Tuscan wine remains a distinct possibility, such gluttony would be far harder to keep up when it comes to eels. The last couple of centuries (and the closing decades of the last one in particular) have seen a drastic drop in eel stocks as counted by the numbers reaching European shores. At barely one-twentieth of historical records, the fish is critically endangered. It's unclear if this drying-up – once so inconceivable that, legend has it, some of the catch was fed to pigs – is down to sheer overfishing, to climate change, to some parasite, or to factors yet unknown. The multiplication of dams, which disrupt Europe's "fluvial continuum," has certainly not helped.

The decline, thankfully, appears to have been slowed if not halted. In 2009, the European Union's Eel Regulation kicked in, with strong protections and export bans. A slight increase in numbers could soon be observed – although a stubborn black market endures. Elsewhere, the East Asia Eel Society, a regional grouping of marine scientists, is seeking to weigh in on resource management policies: there too, eel arrivals have greatly shrunk since about 1980. The society's founder, Katsumi Tsukamoto of Japan, is probably the world's foremost authority on freshwater eels: he was the first to collect *Anguilla japonica*'s eggs in the wild and locate, in the process, the spawning grounds of the fish known to the Japanese as *unagi*. With his country the world's largest eel consumer, Tsukamoto has been speaking up for conservation measures and moderating demand.

Deep legacy, reduced presence

KNOW
YOUR FISH

In the world's current state of knowledge – no human has ever witnessed the mating of eels – full-cycle farming of the eel is impossible.
Aquaculture in the European Union, China and Japan provides the bulk of the world's eel supply, but the process consists of growing out captured elvers. As a dish eaten by the forkful, elvers were themselves once popular in European countries: some are still consumed as a delicacy in Spain, where they are known as *angulas*. These run to EUR 1 000 per kilo, and up to five times that for the first catch. Given a consensus that *angulas* lack any real flavour, their purchase by expensive restaurants is arguably more about extracting public relations value than intrinsic gastronomic worth. Consuming baby eels further squeezes the species' commercial availability by slashing the starter pack.

Unlike elvers, adult eels are dense in taste and texture, with a pleasantly slippery mouthfeel and a moreish melding of ocean vigour and riverine sweetness. The fatty flesh, bursting with vitamins A and B12 (100 grams of eel will amply cover the recommended daily allowance) takes well to grilling or cooking in sauces, or to a combination of these techniques.

In the Japanese *kabayaki* method, the fish is sliced transversally, deboned, skewered, then repeatedly shuttled between the open grill and a bath of soy, sugar and mirin rice wine. The glazed, amber-hued fillets are layered over rice and, in the more upmarket *unagiya* (eel bars), served in swish lacquered boxes. Much closer to FAO's Roman headquarters, the female eel, known in Italian as *capitone*, is a traditional Christmas dish, prepared in all imaginable ways. Consuming the *capitone*, whose shape harks back to the serpent in the Christian tradition, is meant to symbolize the triumph of good over evil. It could be that, of course – or else, the simple fact that when it comes to eels, the female is much longer, thicker and more copious than the male.

Nutrition facts

FISH, EEL, MIXED SPECIES, RAW
per 100 grams

ENERGY (kcal)	184
PROTEIN (g)	18.4
CALCIUM (Ca) (Mg)	20
IRON (Fe) (Mg)	0.5
ZINC (Zn) (Mg)	1.6
SELENIUM (Se) (μg)	7
VITAMIN A (RETINOL) (μg)	1 040
VITAMIN D3 (μg)	23
VITAMIN B12 (μg)	3
EPA (g)	0.084
DHA (g)	0.063

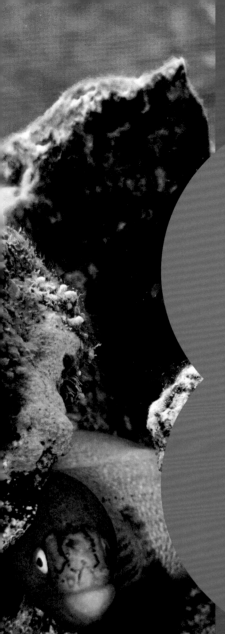

It was very difficult to secure this interview, so I'll skip the pleasantries if you don't mind. Can you confirm that you were born in the Sargasso Sea?

I have been very clear in the pre-interview: there are things about which I will not go on the record, and this is one of them.

Let me put it another way then. Are you aware of the theory, almost universally accepted so far, that you come to life in the Sargasso?

I'm aware that's what the Danish biologist Johannes Schmidt purported to prove in 1920, and that it's become the authoritative view among humans. He spent decades chasing me on the high seas, so I'm grateful for his interest.

But you won't confirm he was right?

I have stated clearly that this is a private matter. As you know, I have not been sighted as an adult in the Sargasso, nor have I ever been seen to spawn. You may also be aware of recent research suggesting that I in fact started life in the Azores.

You're referring to a paper which partly relied on the study of your otolith, the small bone in your head that's supposed to encapsulate your history...

Correct. A century after Schmidt, a team of Chinese-American, French and Japanese scientists found manganese in my otolith, which is present in abundance along the Mid-Atlantic Ridge, but absent in the Sargasso. Beyond that, the standards of scientific evidence required by you humans are not my concern.

I can see we won't get very far with this. Can you at least solve another mystery and tell me what suddenly pushed you to return to the ocean?

I don't know. Can you tell me what pushed you to ditch your job and blow your savings on a sports car?

THE INTERVIEW
EEL

Eel à la Struga

JAGULA NA STRUSHKI NACHIN, ЈАГУЛА НА СТРУШКИ НАЧИН

Red-roofed Struga, on the banks of Lake Ohrid, was once known as Enchalon, the ancient Greek word for "eel". The town's symbolic relationship to the fish remains close: in 2010, the sculptor Sergei Cingulovski honoured the eel by creating a semi-submerged wood installation, snaking along the lake shore for nearly 200 metres. But it's only in the last few years that efforts have begun to create eel passes and restore historic migration routes: starting in the late 1960s, damming of the River Drim, which connects Lake Ohrid to the Adriatic, had severed the fish's ultimate access to the Sargasso Sea.

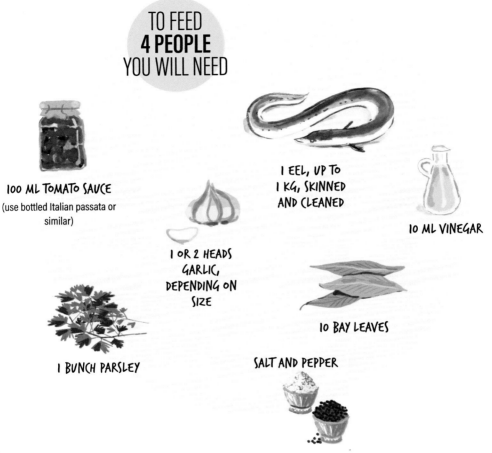

TO FEED
4 PEOPLE
YOU WILL NEED

100 ML TOMATO SAUCE
(use bottled Italian passata or similar)

**I EEL, UP TO
I KG, SKINNED
AND CLEANED**

10 ML VINEGAR

**I OR 2 HEADS
GARLIC,
DEPENDING ON
SIZE**

10 BAY LEAVES

I BUNCH PARSLEY

SALT AND PEPPER

METHOD

1 Divide the eel mentally into two halves lengthwise. Make deep incisions into one side of the fish, all the way through, every 5 centimetres or so. The idea is to keep the eel in one piece but to increase its deployed length, like an origami garland with one continuous and one tasselled side.

2 Line the bottom of a round clay dish with the bay leaves, then place the eel over the leaves and arrange it in a spiral, cut side facing out, so that is covers the surface of the dish. Peel and slightly crush the garlic and stick one clove between each incision in the eel.

3 Pour 100 ml of water over the eel. Place the dish in a preheated oven, at medium temperature. Bake for 20 minutes to tenderize the fish.

4 Take the pot out of the oven and discard the fatty liquid. Add the chopped parsley, pour in the vinegar and tomato sauce, and season with salt and pepper. Bake for an additional 2 hours at 180 °C. Your eel should come out half soft, half sticky and charred.

Flying Fish

PAREXOCOETUS BRACHYPTERUS

"The last strange fish is the last strange bird / Of him no sage hath even heard," wrote the poet John Gray. Hardly true. First, because the flying fish, thought to have appeared tens of millions of years ago, long predates any sages. Second, because by the time Gray was writing in the 1920s, the sages had thoroughly classified the flying fish family, with its forty or so species. Classification, that said, is not enough to stifle fascination, and this is one thing flying fish continue to exert: their perceived dual nature – half fish, half bird – taps into a mythological universe that includes sirens and centaurs. The ancient Greeks thought flying fish spent the day in the sea and the night on land: their scientific name, Exocoetidae, is Latinized Greek for "outside sleepers". They're known to hurl themselves at boats: the French anti-ship missile Exocet is named after them. In Yann Martel's novel *The Life of Pi*, a providential rain of flying fish keeps the stranded hero and his tiger companion from starvation – and incidentally, keeps the first from being devoured by the second.

Flying fish are decisively fish, not birds – albeit endowed with unique pectoral fins. Deployed much like wings, these allow the fish to glide for 150 metres or more. The fins, though, do not flap to keep the fish airborne over long distances. Nor do they help propel the fish out of the water in the first place. In riding the ocean winds, flying fish may escape predators below, but expose themselves to predators above – trapped between worlds, as it were, not so much supercharged hybrids as vulnerable in-betweeners.

Fins aside, flying fish look and taste much like sardines – the flesh deliciously salty-sweet, moderately oily, and rich in heart- and liver-friendly compounds known as phospholipids. They are found and eaten across tropical and subtropical zones, from Japan – where flying fish roe, *tobiko*, tops a classic sushi roll – to Barbados, where the flying fish is an outright national symbol. Our recipe comes from there.

KNOW
YOUR FISH

Flying fish are shaped rather like straight bananas – tubes that taper at both ends. Around 30 centimetres long (pushing half a metre on occasion), they are grey-silver in colour. The pectoral fin, delicate and semi-translucent, begins where the head ends and runs the length of the body. The eye itself is huge, with the black pupil occupying almost its entire surface. Finally, the tail is unevenly forked, with the lower lobe longer than the top one: it acts as a kick starter for flight.

It's unlikely you'll encounter the roe, *tobiko*, outside of sushi restaurants: you'll recognize it from its red-orange hue. *Tobiko* tastes, once again, both salty and sweet. Alongside its high nutrient value, it contains large amounts of cholesterol, so it's best not to binge on it.

Nutrition facts

FLYING FISH, MUSCLE TISSUE, RAW
per 100 grams

ENERGY (kcal)	96
PROTEIN (g)	21
CALCIUM (Ca) (Mg)	13
IRON (Fe) (Mg)	0.5
ZINC (Zn) (Mg)	0.8
SELENIUM (Se) (μg)	0
VITAMIN A (RETINOL) (μg)	3
VITAMIN D3 (μg)	2
VITAMIN B12 (μg)	3.3
OMEGA-3 PUFAS (g)	0.2
EPA (g)	25
DHA (g)	0.15

It's good to meet you. When I told friends I was doing this interview, some weren't aware you existed in real life.

Likewise. All I can say is people should get out more. I, for one, could never afford to doubt the existence of humans. But I'm pleased that you think of me as a bit of a fairy tale character.

Well, you know what they say about humankind. We always wanted to fly; it's wired into our psyche. And we did eventually fly, though not organically. But I digress. What do things look like from up there?

You make it sound as though I soar into the sky. But I don't fly that high. And it's not even technically flight, as you know. I just ride the wind currents for brief periods.

One of the great writers of the twentieth century, E. M. Forster, said that English literature is a flying fish. Do you know what he meant by that?

You'll have to forgive me.

He meant that literature expresses the beauty concealed in an inhospitable world, as you do for the sea.

Well, that's very flattering. It's true, the sea is hostile territory. If it weren't, I wouldn't need to fly. Home though it is, the sea is a place of peril and menace. Tuna, swordfish, squid – they're all after me. So you see, unlike you, I don't fly for pleasure. I fly for my life.

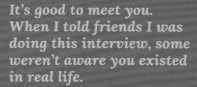

THE INTERVIEW
FLYING FISH

Cou–cou and flying fish

If Barbados has relinquished its status as a monarchy by becoming the world's youngest republic in 2021, other aspects of nationhood, such as the attachment to flying fish, appear more entrenched. The animal features large on Barbadian – or, more colloquially, Bajan – passports, on silver dollar coins and in the collective psyche. But climate change and increasingly frequent invasions of the brown sargassum weed are driving flying fish further and further out from the island nation's shores. As a result, hauls are fewer and more meagre, and the fish, once a poor Bajan's staple, ever more costly. (Amberjack, known here as amber fish, is emerging as an alternative.) Through its CC4FISH project – short for Climate Change Adaptation in the Eastern Caribbean Fisheries Sector –, FAO is working with the Barbadian government to lessen the blow to fisherfolk from warming seas. Satellite data and high-tech sensors are meanwhile being rolled out to give early warnings of incoming sargassum.

Cou-cou is a mix of cornmeal (or sometimes breadfruit) and okra. The cornmeal should be available prepackaged from Caribbean food stores – if not, use polenta from Italian or Balkan delis, or from large supermarkets in many countries. US grits will also do. In combination with flying fish, cou-cou is the dish closest to Bajan hearts. Some punchier versions involve adding lime juice, allspice and Scotch bonnets to the sauce.

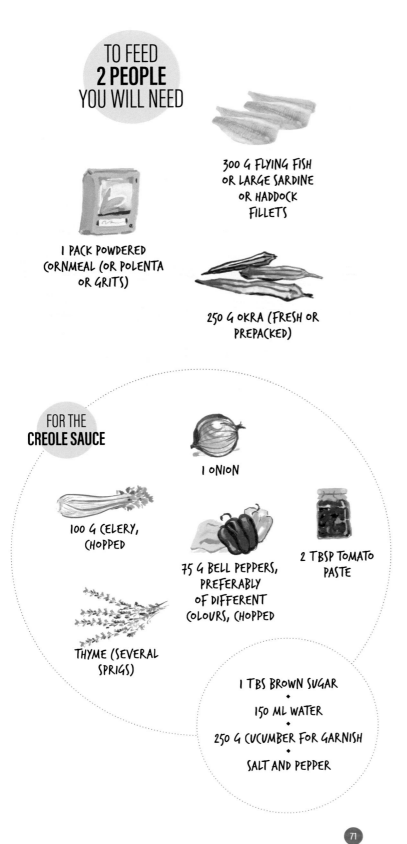

TO FEED
2 PEOPLE
YOU WILL NEED

300 G FLYING FISH
OR LARGE SARDINE
OR HADDOCK
FILLETS

I PACK POWDERED
CORNMEAL (OR POLENTA
OR GRITS)

250 G OKRA (FRESH OR
PREPACKED)

FOR THE
CREOLE SAUCE

1 ONION

100 G CELERY,
CHOPPED

75 G BELL PEPPERS,
PREFERABLY
OF DIFFERENT
COLOURS, CHOPPED

2 TBSP TOMATO
PASTE

THYME (SEVERAL
SPRIGS)

1 TBS BROWN SUGAR
·
150 ML WATER
·
250 G CUCUMBER FOR GARNISH
·
SALT AND PEPPER

METHOD

1   Make your stew base: pour the oil into a large frying pan and sauté gently the chopped onion, garlic, peppers and thyme for 10 minutes or so, or until the vegetables are soft and fragrant. (Include the crushed allspice and a sliver of Scotch bonnet if desired.)

2   Add the chopped tomatoes, tomato paste, sugar and water and leave to simmer for 20 minutes. Season with salt and pepper. Add more water as needed to ensure a loose sauce.

3   Rub the fish fillets in lime juice (optional), roll them up and place them delicately in the sauce. Simmer further till the fish is tender and the sauce has thickened somewhat.

4   While the fish is cooking, boil 400 ml of water in a kettle or in a saucepan on the hob. Add a good pinch of salt. When the water is bubbling, pour in the cornmeal and stir energetically into a dense mush. (Keep the flame low and keep your distance – the cornmeal may spit and scald you badly otherwise.) Add the okra and stir for another minute.

5   Scoop the cornmeal and okra cou-cou into a serving bowl, top with the fish and sauce, and grate your fresh cucumber over it.

Grouper

EPINEPHELUS, MYCTEROPERCA

....

To borrow the sermonizing old dictum about children being seen and not heard, groupers (Epinephelinae) should perhaps be eaten and not seen. Looming around the world's rocky and coral reefs, their bodies are misshapen lumps. The head is indistinct. Out of lateral bulges, cold eyes stare out. So wide are the jaws that they might be the animal's whole *raison d'être*. A grouper's mouth is like a cauldron: with its triple row of spear-like teeth, it could stand in for the gates of hell. More dental plates line the pharynx. As groupers tend to swallow their prey whole, they like to ensure it gets nicely crushed on its way down.

To complete the tableau, we must mention size. This tends to vary across the 160 or so grouper species – as do the colours, which range from brightest red to dullest grey, via every manner of stripe and speckle. But groupers are rarely less than chunky fish. The Atlantic goliath (*Epinephelus itajara*) can grow to 2.5 metres. The black grouper (*Mycteroperca bonaci*), whose habitat extends from the Caribbean to southern Brazil, rocks up in second place, with 2.3 metres. The giant – or "Queensland" – grouper of the Indo-Pacific (*Epinephelus lanceolatus*) comes in at 1.8 metres. And so it goes, down to species that we might more readily imagine handling in the kitchen.

Behaviour-wise, groupers are protogynous hermaphrodites: they start out as females and, over the course of their lives, change to male. This process remains a little mysterious. But we know it tends to be socially mediated, in that a female will transition to male when males are underrepresented within a certain area. This, one might argue, sets new standards for the pursuit of gender parity. Then again, a male grouper will have a harem of females at his disposal – so perhaps it's to do with patriarchal self-interest after all.

If groupers fail the "cute animal" test by a sea mile, they do – true to nature's occasionally perverse compensation games – make up for it in taste. Their flesh is sweet, low in fat and thalassic in a genteel kind of way, all with a moist, flaky texture. Or could it be that groupers' unsightliness evolved as a protective feature, a decoy for their succulence? If so, it didn't work – at least, not with humans. Many grouper species are overfished. The goliath is critically endangered: at the time of writing, it was still protected in the United States of America and the Caribbean by moratoriums brought in 30 years ago.

KNOW
YOUR FISH

Method aside, the answer as to whether your grouper is sustainably fished will lie at the intersection of species and zone. Black grouper from the Gulf of Mexico or south Atlantic, for example, would be a safe choice. By contrast, the Warsaw grouper (*Hyporthodus nigritus*) from much the same waters would be near-threatened, and therefore a more fraught option. Selling the goliath in US or Caribbean markets would be, for now at least, downright illegal. One ambitious 2020 study led by the Marine Science Institute at the University of Texas, which set out to reassess all grouper species around the world, concluded that over a quarter of these were threatened in the wild.

Grouper, that said, is successfully – and often sustainably – cultured, mostly in East Asia and Southeast Asia. Production tends to be seasonal, as demand shoots up around Lunar New Year. Farming grouper, however, is a delicate job. Big and ungainly as they may be, groupers are sensitive animals: a high susceptibility to stress increases hatchlings' vulnerability to disease. In addition to imposing strict hygienic measures, the more advanced hatcheries have been experimenting with tank shape and colours to reduce stress levels. When buying grouper, look for flesh that is firm, springy and moist (as well as the usual absence of odour). If you're getting fillets, the pieces should be creamy white to soft pink, with no browning or discoloration around the edges.

Our grouper recipes come from either side of the Atlantic, in the Caribbean and West Africa: both have instant recognition in their respective countries, and a high lip-smacking index.

Nutrition facts

GROUPER, MIXED SPECIES, RAW
per 100 grams

ENERGY (kcal)	92
PROTEIN (g)	19.4
CALCIUM (Ca) (Mg)	27
IRON (Fe) (Mg)	0.9
ZINC (Zn) (Mg)	0.5
SELENIUM (Se) (μg)	37
VITAMIN A (RETINOL) (μg)	43
VITAMIN B12 (μg)	0.6
EPA (g)	0.027
DHA (g)	0.22

I hope you don't mind doing this interview remotely. I've never spoken to a goliath grouper before and I have to be honest: meeting you face to face makes me a little uncomfortable. Those teeth...

I frankly doubt that I've been more of a threat to you over the years than vice versa. But I suppose meeting online is now routine for all of us, so I'm willing to overlook the offence.

Well, you wouldn't fit through my front door in any case. But let's move on. Your name in English – grouper – doesn't seem appropriate. You're not comfortable in groups: from what I know, you're a largely solitary fish.

You're guilty of what's known as folk etymology. My name in English has nothing to do with "group". It comes from the Portuguese *garoupa*, which itself is thought to be a corruption of an original South American word. Read up before you interview guests: it's basic courtesy.

I stand corrected. But I also find you a little aggressive.

As long as you're not a crustacean, or an octopus, or even a small shark, you're fine.

There was one incident in the Florida Keys in the 1950s...

I wouldn't know. I wasn't born yet and I would advise you not to extrapolate.

All right. I fear we've started off on the wrong foot (not that you have any), so thank you for your time. I have another call booked anyway.

I would normally roll my eyes at this further incivility – but they're so small you wouldn't notice. Goodbye.

THE INTERVIEW
GROUPER

75

South Andros berl fish

The Bahamas may number just 400 000 people, but the span of its islands gives it a diversity of culinary traditions associated with much larger countries – and a similar degree of *terroir* awareness. Technically composed of three islands but forming a single territorial unit, Andros is big on space, low on residents, and altogether quite special. Its ecosystem includes one of the world's longest reefs.

Logically enough in this grouper-friendly environment, "berl" (boil) fish hails from here – a sticky, soupy stew that uses the pieces of fish you might usually discard: in them lies its deliciousness. By the same logic of geographic precision, Bahamians will insist that their onions come from the island of Exuma, reputed for its peppery alliums. You will not get those outside the Bahamas, but you may be able to approximate the flavour by adding freshly crushed black peppercorns.

The fish itself, traditionally grouper, can be substituted with snapper, sea bass or cod: just make sure it's a decent size. (You can save the fillets for another recipe.) Berl fish would be served in the Bahamas with johnny bread, a sweet-savoury cornmeal bake that's halfway to a biscuit, or else – the faster and easier option – with grits or polenta.

TO FEED
2-3 PEOPLE
YOU WILL NEED

1.5 KG GROUPER
(or other quality white fish) head,
fin and tail, all cut up

1 LARGE ONION

2 LARGE LIMES

**1 DESEEDED CHILLI PEPPER
FOR A MILDLY SPICY DISH,**
or anything up to 5 "finger" peppers
for a scorching Caribbean Scoville
count, chopped

**3 STRIPS SALT
PORK OR BACON,
CUT INTO 2.5CM
PIECES**

**2 LARGE POTATOES, NOT
TOO STARCHY, CUT INTO
LARGE CHUNKS**

**SEA SALT AND
FRESHLY CRUSHED
•
BLACK
PEPPERCORNS**

**4 TBSP
UNSALTED BUTTER**

TO MAKE
**THE JOHNNY
BREAD**

800 G FLOUR

4 EGGS

**250 ML SEMI-
SKIMMED MILK**

**30 ML
VEGETABLE OIL**

**60 G UNSALTED
BUTTER +
30 G MELTED
BUTTER**

90 G SUGAR

1 TSP SALT

**6 TBSP BAKING
POWDER
•
125 ML WATER**

METHOD

1 Squeeze one lime over the fish pieces and rub them with sea salt, then wash them well under cold running water until the juices run clear.

2 In a heavy-bottomed pot, sweat the onions, potatoes, bacon and chilli gently with the butter, so that they're coated in melted fat, but not sticky or burned.

3 Season with sea salt and crushed pepper and add the fish. Pour in just enough water to cover the fish halfway. This will ensure a savoury broth. Cover with a lid and simmer for 12 minutes or so, checking the potatoes with a fork for softness. Squeeze in the remaining lime, adjust for seasoning, and you're done.

FOR THE JOHNNY BREAD

1 Preheat the oven to 180° C.

2 Sift the flour and baking powder into a mixing bowl. Add the sugar and salt, then the butter in small chunks. Knead it all in.

3 Now add the eggs, oil, water and milk. Mix everything into a batter. The mix should be softer than if you were baking bread, but harder than if you were baking cake, so aim for a hybrid. Moisten further as needed.

4 Grease an oven tin, pour the batter in and bake for about half an hour. When it's still fairly soft, remove from the oven, brush the mix with the melted butter, and bake some more until the top is firm and golden. Rest the johnny cake for a few minutes before turning out and serving.

Thieboudienne

Thieboudienne, which you may find spelled in a variety of ways – it's pronounced tchebou-DJEN – stands for "fish and rice" in Wolof. It is to Senegal what couscous is to the countries of the Maghreb. On the face of it, this national dish is a straightforward mix of protein, vegetables and starch. But it's a work of many steps; it comes with a stack of lateral ingredients; and such is its sense of festive informality, and so deeply exuberant its flavours, that you'll want to live inside it.

Thieboudienne is historically associated with Saint-Louis, in northern Senegal. A political and economic centre in colonial times, the city has long ceded pre-eminence on all fronts to Dakar. But it remains much loved for its UNESCO-listed architecture and languid ocean feel. Its fisherfolk, though, are being squeezed by dwindling stocks and rising sea levels. You'll often find thieboudienne made with chicken or beef these days, either under the influence of couscous, or just because there's less fish and more meat to go around than in the old days. Our recipe, as per this book's remit, sticks to the original. The

type of fish is not prescriptive – but it should be a white ocean fish.

You'll also need a handful of dried *bissap blanc* (white hibiscus flowers), which can be sourced online or from West African grocery shops; and a piece of *guedj*, a smoked salted fish used as a condiment: substitute other strong cured fish, or even a couple of anchovies. Finally, a full traditional rendition of thieboudienne would include *yet*,

a dried, fermented sea snail added for pungency, but this you can omit or replace. There are, in fact, as many versions of thieboudienne as there are spellings. Some feature aubergine; ours doesn't, but is otherwise fairly complex. Certain seasonings are split into two, with one half cooked longer than the other for added layering of flavour. Feel free to simplify things depending on availability – that of ingredients, and your own.

TO FEED 6-8 PEOPLE YOU WILL NEED

2 FISH
(grouper or other) of around 500 g
each, heads off, cleaned

1 ½ KG JASMINE RICE

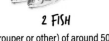

**2 TURNIPS, 2 CARROTS, 1 MEDIUM
WHITE CABBAGE, CHOPPED LARGE**

**1 PIECE MANIOC,
ABOUT 10 CM,**
pre-soaked and
peeled, sliced thickly

**1 HANDFUL
CHERRY TOMATOES**

3 ONIONS
sliced lengthwise,
3 SPRING ONIONS (SCALLIONS)

**1 SMALL SQUASH,
PUMPKIN, GOURD,
MARROW OR SIMILAR**

**1 HANDFUL
OKRAS**

**1 BUNDLE DRIED
WHITE HIBISCUS
(TIED)**

3 GARLIC CLOVES

1 PIECE FERMENTED SEA SNAIL
(optional - substitute 1 tbsp Vietnamese
fish sauce or 1 tsp miso paste if desired)

**1 PIECE STRONG DRIED
FISH FOR SEASONING**
(or 2-3 cured anchovies)

**1 TBSP OR LESS
POWDERED CHILLI,
TO TASTE**

**300 G DOUBLE TOMATO
CONCENTRATE**

**100 ML
SOAKED
TAMARIND
FLESH**

**1 TSP
SHRIMP PASTE
FROM A JAR**
(optional)

1 BUNDLE PARSLEY
•
2 HABANERO
or other chillies, deseeded, chopped
•
1 TSP SUGAR
•
SALT AND PEPPER
•
**VEGETABLE OR PEANUT OIL
FOR FRYING**

METHOD

1. First, make the stuffing for the fish. In a mixer, blend the parsley with the spring onions, garlic cloves, dried chilli, salt and pepper. Season and stuff the fish, rubbing any leftover parsley mix over them, then slice them into 2.5-cm pieces through the bone.

2. Fry the fish pieces in a large pan with a little oil, in batches if needed. The fish should be golden on the outside, but not cooked through. Drain the excess oil.

3. Separately, heat more oil in a large saucepan. Add the sliced onions, one of the chopped habaneros, the sea snail (or fish sauce or miso), and a little salt. Cook, stirring, for 5 minutes, until the onion starts to caramelize. Now add the tomato concentrate and cook 10–15 minutes until the paste begins to form lumps. Add a little water to keep from burning.

4. Tip in the cherry tomatoes and keep stirring for 5 minutes, add 3.5 litres of water and bring to the boil. Season and add the carrots, turnips, manioc, cabbage, half the dried fish or anchovies, the bundle of hibiscus and the soaked tamarind flesh, and leave to bubble gently for 15 minutes.

5. Place the fish pieces, the okras and the remaining chopped habanero into the pot and cook gently for a further 10 minutes, then remove fish and vegetables to a large bowl. Cover with cling film and keep warm.

6. Cook your rice two-thirds of the way in a rice cooker or on the hob, adding the remaining half of the dried fish or anchovies. When the rice has absorbed the water, but retains a crunch, remove 3 or 4 ladles of the sauce, then place the rice in the saucepan where most of the sauce still is. Finish cooking the rice in the sauce, so that you end up with rice that's rich and slick, but not soupy.

7. In a milk pan, heat up the small quantity of sauce removed from the main pot, dissolving the shrimp paste in it, and adding the sugar. Plate up the rice onto a tray, and spoon the warmed-up sauce over it. Arrange the vegetables over the rice, and place the fish steaks on top to serve.

Herring

CLUPEA HARENGUS

Even as the Cold War was winding down in the 1980s, suspicious subaquatic noises in the waters off southern Sweden suggested persistent enemy submarine activity. Soviet incursions were nothing new. But this time, no detection system could pick up any vessels. For a decade and a half, the mysterious acoustic phenomena remained an irritant in relations between Stockholm and Moscow. "It sounded like someone frying bacon," said Marcus Wahlberg, an academic who, in 1996, was brought in to study the bubbling sounds. He was the first civilian to hear them – and identify them. They weren't submarine noises: they were herring farts.

Herrings are oily, cold-ocean creatures, with dozens of Atlantic and Pacific subspecies. Like most fish other than sharks and fellow cartilaginous species, they have swim bladders. These function like internal balloons: by storing or releasing gas, they help regulate the creature's buoyancy. Herrings, however, have one unique feature: their swim bladder connects directly to their anus. Gas is also expelled under stress – for example, when schools, which may count millions of individuals, brush past potential predators. It was, quite simply, herring flatulence that stumped the Swedish defence establishment. The herrings' sudden presence in the area exemplified a pattern of abrupt appearances and disappearances that has spurred talk of miracles through the ages. One theory has it that as a new age cohort matures and becomes dominant in a herring group, it sets the direction of travel. How the cohort communicates this, assuming the hypothesis is correct, remains unknown. Then again, almost everything about how fish communicate – and they do, amply – has yet to be revealed.

A working-class fish

Moving as it does in million-strong packs, the herring resembles its relative, the anchovy. Gastronomically and culturally too, the analogy stands: the former is to the north what the latter is to the Mediterranean. Economically, the herring, somewhat larger than the anchovy and richer in fat, has been even more significant. A staple source of calories throughout the upper part of the northern hemisphere, it was fiercely fought over by the English (later British) and Dutch navies in the seventeenth and eighteenth centuries. And unlike the anchovy, which has gentrified in recent decades, herring has remained a people's fish. Take the French *harengs pommes à l'huile*, which consists of just that: herring fillets, boiled potatoes and warm oil, plus raw onions. The most you'd get as a bourgeois concession is a little parsley. In Germany, folk wisdom associates the *rollmop*, where the pickled fillet is tightly rolled around a gherkin, with hangover cures. Over in Poland, the traditional way of serving herring is just lightly salted, or perhaps with cream.

While such austerity may seem quaint, it still tends to define herring. The lack of accoutrement may reflect an empirical understanding that the fish is densely nutritious in its own right, an onslaught of vitamin D and fatty acids. In Alaska, and in what is now Western Canada and the northwestern US states, First Nation tribes rendered herring for its oil. So did, an ocean and a continent away, the Swedes, who also burned the oil for lighting and used it to fertilize the land. On the plate, because of herring's oleaginous intensity, garnishes that work best are either starchy flavour absorbers such as potatoes, or else sharp, acidic competitors. In northern Europe, accompaniments may involve tomato,

horseradish or *hovmästarsås*, a dill and mustard sauce. Sour apple is another great pairing, and so is rémoulade, a loose mustardy mayonnaise with shredded celeriac and chopped gherkins.

Few fish are quite as versatile as herring: it can be grilled, pickled, marinated, smoked and so forth. Beyond the diversity of experience, there's a health aspect to this extent of processing. Herrings are prone to nematodes, a type of parasitical roundworm that can cause anaemia and debilitating gastro-intestinal illness in humans. These days, nematodes are killed by freezing the fish to 45° C below zero; in the past, the job was done by brining or other cures, or by heat treatments. The most notorious method may be the one involved in *surströmming*, a months-long fermentation treatment perfected in (yet again) Sweden: the result is, to many outsiders, evocative of a putrid stink bomb.

Versatility out of necessity

Oddly enough, it's the one upper-class incarnation of herring, as cold-smoked breakfast kippers for the English gentry, that involved only light brining and was thus insufficient to eradicate the roundworm. Or, who knows, it may have been a subconscious way to invite the unforeseen into the regulated setting of the stately home – an aristocratic form of dicing with discomfort.

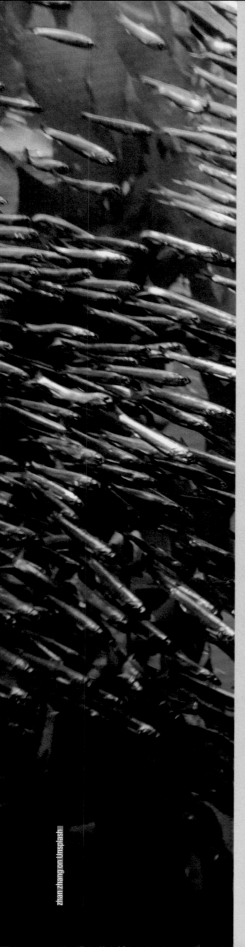

KNOW YOUR FISH

Herrings are pretty, elongated fish, shimmering silver in colour. (Some intense, traditional salt cures are said to make the fish glow almost phosphorescent at night.) Adults may measure up to 35 centimetres or so, though are commonly found at a little over half that.

Herring remains plentiful and cheap: it has long been used as bait for more valuable species, including the highly profitable lobster. That in itself means herring is in demand. In January 2022, US federal officials charged five fishers from Maine and one from New Hampshire with conspiracy, mail fraud and obstruction of justice for failing to report more than 1 000 tonnes of herring landings. The fish is alleged to have been sold directly to fish dealers and lobster boat operators.

So while your herring is unlikely to be the object of fish fraud as such (no one would fake a herring; at most, it might get mixed up with its fellow clupeid, the sardine), its affordability is no ironclad guarantee that it's been legally fished. Here as in other cases, it pays to buy from reputable outlets, with full traceability. And if you're getting your herring fresh, the usual criteria apply: bright unblemished skin, neat gills, clear eyes and absence of blood or slime. Fatty cold-water fish spoils first and smells worst, so when it comes to herring, the alert system is unambiguous.

ATLANTIC HERRING FILLET (WILD)
per 100 grams

PROTEIN (g)	17.9
IRON (Fe) (Mg)	1.1
ZINC (Zn) (Mg)	0.7
IODINE (I) (μg)	24
SELENIUM (Se) (μg)	38
VITAMIN A (RETINOL) (μg)	36
VITAMIN D3 (μg)	30
VITAMIN B12 (μg)	12
OMEGA-3 PUFAS (g)	1.73
EPA (g)	0.548
DHA (g)	0.71

Where did you get your name from, if you don't mind me asking?

Hard to tell. In Scandinavian languages, but also in Finnish and Russian, I'm called *sild* or versions of it. The word is derived from Old Norse, but beyond that, no one is sure. In other Germanic and most European languages, including English, there's a theory that my name comes from *heri*, or "army" (*Heer* in modern German), arguably because I move in huge, regimented schools.

Indeed, there's a lot of you around, which tends to make you inexpensive.

True, in Europe they don't think of me as a premium food fish. But elsewhere I'm held in higher regard. Egyptians, who import me from northern Europe, have made me part of *fissekh*, a festive meal consisting of smoked and fermented fish. It's eaten for Sham el-Nassim, a Pharoah-era springtime ritual that sees people visiting farms and gardens. Also, the Nuu-chah-nulth, a Native American group, used to give me equal billing with salmon: they believed the salmon and I lived in side-by-side underwater houses.

Did you?

No. Given salmon's predatory ways, I doubt we'd have had good neighbourly relations. I would have been terrified to venture outside.

Fair enough. On your previous point – I understand you're also considered a bit of a delicacy in Japan?

My roe is. It's called *kazunoku*, and it's eaten for New Year. It's golden and crunchy – the Japanese refer to this texture onomatopoeically as *kori-kori*. It looks like glorious pear segments, in fact. Try it if you get a chance.

THE INTERVIEW
HERRING

Soused herring

MAATJESHARING

Soused herring is closely associated with *Vlaggetjesdag* (Flag Day) in mid-June, when tens of thousands of people mob the fish stands of Scheveningen, The Hague's beach suburb. There are residual religious overtones to the festival, but it mostly involves herring-themed revelry. Participants chomp on their fish in the least refined way possible, by holding it up by the tail, high above their heads, and lowering it into their mouths, studded with bits of raw onion.

Maatjesharing is literally "maiden herring". The fish must be young, caught between late May and early June, when it's just starting to put on some fat after a winter diet of scarce plankton pickings. (The first catch of the season is auctioned off and the proceeds donated to charity.) At this stage of development, the herring is free of milt or roe, and its taste less pronounced. The fish is first "gibbed" – a gutting method perfected in the Low Countries in the Middle Ages. The gills

and some of the innards are removed, but not the liver or pancreas: these go on releasing enzymes that round off the flavour. "Sousing" refers to a mild brining cure that hovers between savoury and sweet: it can consist of salt, vinegar and sugar, or else the vinegar could be replaced with cider, or even, on occasion, with tea.

Our recipe is a modified version of what you might get on V*laggetjesdag*, but you may vary things to taste – by adding capers, for example, and sour cream or mayonnaise. Prepare this a day or two in advance.

APPETIZER FOR
2 PEOPLE
YOU WILL NEED

4 YOUNG HERRING FILLETS (DOUBLE THE QUANTITIES IF SERVING AS A MAIN COURSE)

I GREEN APPLE, GRANNY SMITH OR SIMILAR

I WHITE ONION

FOR THE
MARINADE

I TSP SUGAR

IOO CL WHITE WINE, OR CIDER, OR DILUTED LEMON JUICE

(you can also combine all three; some recipes suggest a splash of vodka)

I GENEROUS PINCH SALT

5-6 CRUSHED PEPPERCORNS

I BAY LEAF

METHOD

1 Combine all marinade ingredients, making sure the sugar is dissolved. Place the fish fillets in the marinade, cover with cling film, and leave in the fridge overnight or longer.

2 When it's almost time to serve, take the fish out of the marinade. Reduce the marinade by about half over a low flame, then strain.

3 Dice the onion and green apple. Plate up the herring fillets, sprinkle them with the diced apple and onion, and pour the warm strained marinade on top. Serve with boiled potatoes, and maybe parsley or chervil.

Mackerel

SCOMBER SCOMBRUS
~ JAPONICUS, COLIAS

Mackerel is a generic designation for some thirty species, of which seven are considered "true" and the rest close relatives. Overall, they tend to prefer company, hanging out in large schools and spawning in cold coastal waters, with vast populations found along north Atlantic shores and in the East China Sea. Italians, who call the mackerel *sgombro*, and Russians (*skumbriya*) will recognize mackerel's Latin family name of Scombridae – a melodious moniker it shares with tuna and bonito.

Smallish and pleasingly torpedo-shaped, true mackerel – but most of their cousins too – have shimmering skin that runs to silvery blues. They sport specks or stripes which, depending on species and perspective, may channel either a marine zebra or the cuneiform alphabet. But if the looks of mackerel are an easy delight, the flavour eschews blandness: bold and downright fishy, mackerel – and Atlantic mackerel in particular – conjures the ocean at its most bracingly metallic. Such is mackerel's reputation for pungency that in the folk tales of Japan (where it is often cured with sugar to tame its assertiveness), the fish is said to repel the tengu, a long-nosed demon sprite. Each time a child went missing, assumed to be taken by a tengu, villagers would roam the forest shouting that the child had eaten mackerel (*saba*) so that the goblin would release it.

Oily delights

As other pelagic fish (meaning those that hug the continental shelf, up to a depth of some 200 metres) mackerel has been used extensively to produce oil and fishmeal. But recent decades have seen it increasingly marketed for direct human consumption. With good reason: brimming with omega-3 fatty acids, mackerel has been classified by FAO as a nutritionally high-value product. Far from greasy despite its high oil content, the fish feels rich and dense on the palate. It also freezes well and lends

itself to all manner of treatments. Fry it, grill it, roast it, smoke it. Pair it with raw acidic flavours, such as diced green apple or vinegared beetroot. Serve it with lightly boiled eggs. Or poach it and toss it into a salad with some spinach leaves: the bitter burst of chlorophyll is a fine match for its tangy punch. (One of FAO's fisheries officials swears by Italian chub mackerel fillets, straight out of the tin and seasoned with olive oil, soy sauce and turmeric powder.)

Reassuringly, mackerel is still plentiful in the world's seas and oceans. Asia and northwestern Europe supply most of the world's catch: in 2020, China and Norway tied for top exporter, followed by Japan, the Netherlands and Denmark. And while the fish remains eminently affordable, its growing popularity has given a fillip to fisherfolk incomes. In Norway, for example, electronic mackerel auctions automatically allocate orders to the highest bidder. Finally, to add to all of mackerel's benefits culinary, nutritional and social, a time-saving grace: it has no scales, or almost.

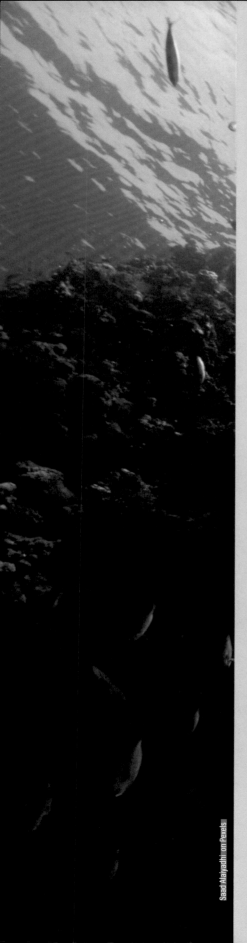

KNOW
YOUR FISH

Easy on the wallet and distinctive in looks, mackerel is not among the fish most prone to mislabelling or fraud: there is little incentive or visual wriggle room to swap inferior fish for it. Look for a silver-blue sheen and tiger stripes or polka dots.

True enough, telling apart certain species can be tricky: the job may require a marine biologist's eye when the fish is whole; dissection otherwise; and in some cases DNA sequencing. Don't let this worry you. Although the four recipes featured here illustrate different traditions – two Asian, two African – and thus call for different varieties, this has a lot to do with what is locally available and less with any intrinsic qualities of the fish. For most recipes, in fact, any mackerel will do. And if you like mackerel, you will like all sorts.

Do pay attention to freshness. Mackerel spoils fast. If buying the fish whole, look for bright round eyes, like newly minted coins. The gills should be crimson. At the first sign of cloudy muck forming over the eye, general sogginess, or skin that is faded or dull, move on. Filleted fish should have flesh that is pale pink, spotless and blood-free.

Nutrition facts

ATLANTIC HORSE MACKEREL, WILD, FILLET W/O SKIN, RAW (N.S.)
per 100 grams

ENERGY (kcal)	123
PROTEIN (g)	18.6
CALCIUM (Ca) (Mg)	30.6
IRON (Fe) (Mg)	1.1
ZINC (Zn) (Mg)	0.4
IODINE (I) (µg)	29
SELENIUM (Se) (µg)	53
VITAMIN A (RETINOL) (µg)	4
VITAMIN D3 (µg)	27
VITAMIN B12 (µg)	6.8
OMEGA-3 PUFAS (g)	1.22
EPA (g)	0.38
DHA (g)	0.84

Those blue highlights – they really suit you.
Thank you. I had them done on my back, so I can't really see them myself. Still, I can tell they turn heads.

But don't you feel rather exposed, going around scale-less?
A little, but I don't mind showing some skin.

Well, you are living proof that one can be both fatty and very handsome. A blow against negative stereotypes.
Yes, it's very affirming. It's also good that people can now see beyond my oil – that they appreciate me holistically, for who I am. Things have been changing for the better.

My editor, who believes in a more confrontational interviewing style, will probably find this conversation rather indulgent. In any case, I'm glad you agreed to talk to me on behalf of all mackerel. Your family is quite diverse, and there's the added complication of shading off into tuna. It matters that you have legitimacy as a spokesfish.
We are a fairly diverse family, but the core of it is quite tight-knit. It's true, there are also some faux-relatives, plus, as you say, a bit of over-lap at the biological edges. Tuna can actually be extremely antago-nistic and imperious – always trying to swallow us as more vulnerable clan members. So I wouldn't say our links are close. But I'm told you'll also be interviewing tuna: I'm sure you can form your own opinion.

THE INTERVIEW
MACKEREL

Cured mackerel

SHIME SABA, しめ鯖

With some 400 000 tonnes in 2020, Japan is one of the world's top mackerel producers, supplying both regional and African markets. Domestic demand too remains strong, absorbing around a third of the output.

Fittingly in a culture that prizes the visuals of food, sashimi pieces sold at counters in Japan are sometimes referred to as *hikarimono*, or shiny things. This elegantly spare recipe, while not using raw fish, requires no heat cooking; it preserves the skin of the fish to create a burst of iridescence on the plate. Cured mackerel would normally form part of a wider dining experience. Serve it as an appetizer, perhaps. Or, if you intend to make this a single-course meal, adjust the quantities to suit.

The variety used here is the "big-eyed" chub Pacific mackerel, measuring around 30 centimetres or a bit more in length. Rice vinegar is available from many supermarkets or speciality Asian food stores: you may substitute apple cider vinegar to replicate its mild flavour. Kelp is a thick, ropelike seaweed. One of the main ingredients in dashi, the savoury broth that acts as the base note of Japanese gastronomy, kelp is usually sold desiccated: by this stage, it resembles a leather strap or piece of tree bark. Replace the kelp with other seaweed if unavailable, or omit it if you must. Its absence will render your marinade less complex – and, incidentally, less rich in vitamins K and B9, which are good for your blood – but your dish won't suffer irredeemably.

TO FEED
2 PEOPLE
YOU WILL NEED

I WHOLE MACKEREL
(LOOK FOR A LARGE,
PLUMP FISH) OR 2 FILLETS

200 G SUGAR
+ ANOTHER 3-4 TBSP

UP TO 400 ML RICE
VINEGAR

200 G SALT

2 PIECES DRIED KELP,
ROUGHLY THE SIZE
OF THE FISH

METHOD

1 To fillet the fish (if buying it whole), lay it on its side, chop off the head if present, and cut off the tail and dorsal fin with a pair of scissors. Slice a sharp flat knife through the fish lengthwise, working it gradually to separate the flesh neatly from the backbone. When you have your first fillet, turn the fish over and slice just above the backbone once more to get your second fillet.

Discard the bone. Pin bones should be removed too, by cutting them out in a long strip. (Mackerel flesh is too delicate to withstand the repeated assault of tweezers.)

2 Remove and discard the outer, transparent skin. To do this, grab one corner of the fillet and peel back in one swoop, as if stripping the plastic film off a new mobile phone.

3 Douse each fillet in sugar on both sides, then again in salt, ensuring that it's totally covered. Refrigerate for 30 minutes. The fish will release excess water.

4 Wash off the sugar and salt and dry the fillets on kitchen paper.

5  Mix the remaining clean sugar with the vinegar and kelp in a largish bowl. When the sugar has melted

and the kelp has softened somewhat, soak the fillets in the marinade. Refrigerate again for up to an hour, or until the flesh of the mackerel has turned a paler shade of pink.

6 Take out the fish and pat it dry. Cut the fillets into medium slices, aiming for a thickness of a little under 1 cm. Serve with soy sauce and wasabi, or with salt and lemon or grapefruit.

Braised mackerel with radish

GODEUNG-EO JORIM, 고등어조림

Mackerel tops surveys as the Republic of Korea's favourite fish. It is also a mascot of Busan, the country's second city. Every October, the southern metropolis dedicates a dazzling beach festival to it.

You will likely skip the fireworks when you cook this recipe at home, but your taste buds will be in a festive mood: it has none of the sobriety of the previous dish. Rather than showcase the main ingredient through muted intervention, it floods it with colours and flavours: this approach to *godeung-eo* – as the chub mackerel is known here – is brash, spicy and exuberantly aromatic.

This recipe also (marginally) uses kelp – though as in the Japanese dish, you can substitute another strong seaweed. The use of persimmon is a modern twist on the classic recourse to sugar: the fruit is rich in antioxidants, and in vitamins A and C. If no persimmon juice is available, make do with mango juice, adding a dash of lemon to dial up the astringency.

The radish referenced here is the sturdy, white-and-pale-green daikon (or, to the Koreans, *mu*), common in most Asian food stores. If none can be had, try turnip instead.

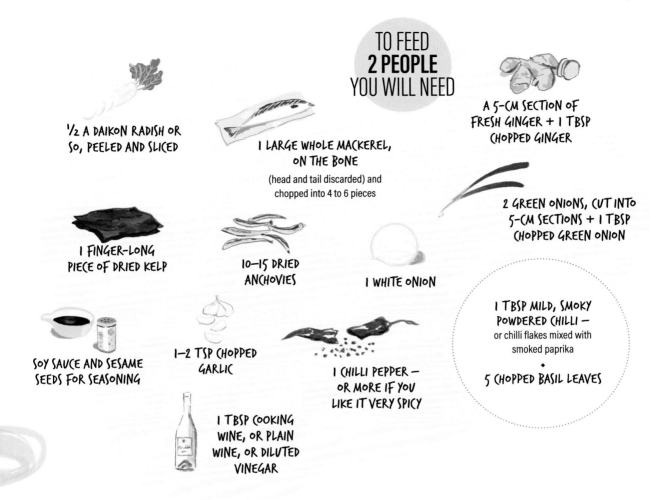

TO FEED
2 PEOPLE
YOU WILL NEED

½ A DAIKON RADISH OR
SO, PEELED AND SLICED

I LARGE WHOLE MACKEREL,
ON THE BONE
(head and tail discarded) and
chopped into 4 to 6 pieces

A 5-CM SECTION OF
FRESH GINGER + I TBSP
CHOPPED GINGER

I FINGER-LONG
PIECE OF DRIED KELP

10-15 DRIED
ANCHOVIES

I WHITE ONION

2 GREEN ONIONS, CUT INTO
5-CM SECTIONS + I TBSP
CHOPPED GREEN ONION

SOY SAUCE AND SESAME
SEEDS FOR SEASONING

1-2 TSP CHOPPED
GARLIC

I CHILLI PEPPER —
OR MORE IF YOU
LIKE IT VERY SPICY

I TBSP MILD, SMOKY
POWDERED CHILLI —
or chilli flakes mixed with
smoked paprika
•
5 CHOPPED BASIL LEAVES

I TBSP COOKING
WINE, OR PLAIN
WINE, OR DILUTED
VINEGAR

METHOD

1 Bring the dried anchovies and the kelp to the boil in 700 ml of water. Turn the heat down and boil the liquid for a further 5 minutes to make half a litre of pungent stock.

2 Take a separate pot and line its bottom with the sliced radish. Pour in a generous splash of soy sauce and the stock. Add the chopped ginger, the sections of green onion and the chilli pepper and bring to the boil.

3 When the liquid has boiled, put in the washed mackerel, the white onion, the chopped ginger, 1 further tbsp of soy sauce, the powdered chilli, the persimmon juice and the cooking wine. Boil down for another 10 minutes.

4 Throw in the chopped green onion and chopped basil leaves. Give the dish another 30 seconds on the boil.

5 Turn off the flame, transfer to a bowl and decorate with the sesame seeds before serving.

Mackerel with moringa leaves

MAQUEREAU AUX FEUILLES DE MORINGA

Way down from Japan or the Republic of Korea in the prosperity league tables, Burkina Faso is landlocked, with a poor record of food security and nutrition. Diets are largely based on agricultural staple crops; beef is the main source of animal protein. Yet here too, urbanization is driving a taste for fish: grill shacks serving the stuff attract large numbers of Burkinabès.

As aquaculture makes modest inroads into the country's arid landscapes, more fish, especially tilapia, is starting to come from domestic sources. Still, the horse mackerel used in this recipe would be imported frozen from Senegal or another of the region's coastal states – or, as is ever more the case in Africa, from Japan.

Moringa (*Moringa oleifeira*) is a plant chock-full of vitamins A and C, calcium and potassium. Its reputation as a malnutrition-buster explains its popularity in South Asia and much of Africa – and increasingly in the west, where it's more commonly sold as a health supplement. The leaves are often described as tasting earthily sharp: from a Western perspective, their flavour falls somewhere between rocket (arugula) and spinach. They are often, though not necessarily, ground up before being added to dishes; in this case, powdered matcha green tea would make a good substitute. If you prefer a grainier, less processed texture – again, assuming moringa is unavailable – use kale, or a mix of kale and spinach, adding some arugula for a tart, bitter kick.

Throw in the arugula at the very end, allowing it to just wilt from the warmth of the dish: do not cook it, or it will turn to inedible mulch.

This recipe, which combines frying and stewing, ingeniously uses two fish. One is served whole, the other flaked and sweated with the other ingredients to form a hearty sauce.

TO FEED
2 PEOPLE
YOU WILL NEED

2 HORSE MACKEREL

3 ONIONS, CHOPPED

4 TOMATOES

**4 SPRIGS PARSLEY,
FINELY CHOPPED**

**1 LARGE HANDFUL MORINGA LEAVES
(SEE ABOVE FOR SUBSTITUTES)**

OLIVE OIL

GARLIC TO TASTE

SALT AND PEPPER

METHOD

1. Fry both mackerel in olive oil. You may want to do this one fish at time to ensure the oil stays hot. Take out, pat the grease away with kitchen paper, and set aside.

2. Boil the moringa leaves in salted water. Remove them with a slotted spoon and leave to cool. When safe to handle, squeeze out the water and reserve the leaves.

3. Tear the flesh off one of the two fried mackerel, and place it in a pan together with the onions, tomatoes, parsley and celery. Add 6 tbsp of olive oil, mix well and warm the mixture up on a medium flame.

4. Add the boiled moringa leaves and season with salt and pepper.

5. Simmer the dish for half an hour or so, until you have a dense, rough sauce. When ready, plate up the whole mackerel with some plain boiled rice, and top with the sauce.

Mackerel stir-fry stew

TIBSI, ጥብሲ.

In this take on the spicy dish known in Tigrinya, the main language of Eritrea, as *tibsi*, fish replaces the more common beef.

Tibsi is itself a faster – but no less rewarding – spin on the long-simmered *tsebhi*, much enjoyed across the Horn of Africa.

Eritrean statistics are published sporadically and can be hard to come by; FAO has no reports of commercial aquaculture being practised in the country. Conversely, its position along the Red Sea – Eritrea is largely a coastal state – gives the nation ample access to waters rich in mackerel, tuna, grouper, snapper, sardines and anchovies.

In its purest version, this recipe requires spiced clarified butter, which is obtained by boiling, and then straining, the butter with chopped onions and ground garlic, turmeric, basil, nutmeg, ginger, pepper, fenugreek, cardamom, cinnamon and cloves. The result is as entrancing as it sounds, and gives Eritrean and Ethiopian cuisines their distinctive aromatic warmth. You're welcome to have a go, or else use plain butter or oil as a frying base and add a reduced combination of spices during the cooking process. The berberè mix, on the other hand, is hard to do without: get it from a specialist shop, or make your own by grinding together chilli, coriander seeds, allspice, nigella and (again) fenugreek.

Bear in mind, all that being said, that in Eritrea *tibsi* is viewed as close to fast food: many of the ingredients would be ready to hand. Elsewhere, you could find yourself sourcing and grinding an array of condiments not once but twice, as well as fiddling with the butter. The staple injera bread, a spongy sourdough pancake, is mouthwatering but takes 2–3 days to ferment. Instructions on how to make it are given below – though unless time is no object, it's clearly more convenient to buy it or replace it with couscous. Altogether, you may decide to cut a number of corners in approaching this recipe. Take heart: even approximating the outcome will be highly satisfying.

TO FEED 2 PEOPLE YOU WILL NEED

1 MACKEREL, WEIGHING ABOUT ½ KG, cleaned and cut into largish bite-sized chunks

1 LARGE TOMATO, FINELY DICED

1 ONION, CHOPPED

2 TSP FINELY MINCED GARLIC

SPICED CLARIFIED BUTTER (see above) or, failing that, butter or vegetable oil

2 TSP BERBERE (see above for making your own if unavailable)

1 TSP BLACK PEPPERCORNS, 1 TSP CUMIN SEEDS AND 1 WHOLE NUTMEG (omit if using spiced butter)

1 DASH SESAME OIL

1 PINCH SALT

FOR THE INJERA BREAD

Injera is made with teff flour, obtained from an ancient grain. It is gluten-free and rich in protein and minerals, and has become popular with endurance athletes beyond Ethiopia and Eritrea, where it is the default bread. Outside the region, you should be able to find the flour in health and organic stores. You will need equal quantities of flour and water, about 4 cups each, plus a pinch of salt.

1. Place the teff flour in a container with a lid, add the water and salt, and mix well to form a smooth batter.

2. Place the lid over the container and leave 2–3 days to ferment, by which time it should have acquired a foamy texture. Add 1 cup of warm water.

3. Let the mixture rest at room temperature for about 2 minutes. It should fizz a little. Meanwhile, spread a clean cloth or kitchen towel on your countertop or work surface.

4. Heat a non-stick frying pan and 1 cup of the batter, spreading it quickly in a circular fashion as if making a pancake.

5. After about 30 seconds, the surface should start to bubble. Cover the frying pan and cook for up to 2 minutes longer, adjusting the temperature to prevent burning. Slide the injera gently out of the pan and repeat with the rest of the batter. The injera can be eaten straight away or kept in the fridge for several days.

METHOD

1. Heat some clarified butter, plain butter or oil in a saucepan and cook the chopped onions and diced tomato until soft.

2. Add the cut mackerel to the pot, with butter or oil if necessary, and brown until hot.

3. Turn the heat down, add the berberè and minced garlic, and continue to simmer until the fish is done.

4. Unless using spiced butter, heat up the peppercorns, cumin seeds and nutmeg in a dry pan until fragrant, taking care not to burn them. Grind these spices together in a mortar and stir into the pot, cooking for another minute. If using the butter, skip this part.

5. Finish with the sesame oil for a good sheen. The *tibsi* may be as dry or wet as you like, so play with adding water, oil or butter as desired through the cooking process. The dish, however, should not be greasy. Serve with salad or fruit, and either couscous or injera (see left for recipe) to mop it up.

Mahimahi

CORYPHAENA HIPPURUS

A young man, slender and naked. He's walking. His skin is the shade of burnt earth. Dangling from each hand, a string of fishes. So goes the *Fisherman*, painted on a wall on the Greek island of Santorini. The fresco was recovered from the West House, a large private residence in the city of Akrotiri, buried under a volcanic eruption some 37 centuries ago – one of the episodes that may have spurred the legend of Atlantis. The young man is now displayed in the Museum of Prehistoric Thera, Santorini's ancient name.

Was the Fisherman really one? Perhaps. But his head is shaved. And on it sit two marine creatures, a sea snail and a squid – all of which points to a symbolic level of representation. It could be that a religious ritual is being depicted; that the young man is solemnly offering the fish to the gods, rather than routinely taking it from the sea.

The fish themselves are mahimahis. We know it from the blue and yellow colouring. We see it in the skin's dotted pattern. We spot it in the blunt head and rictus-shaped mouth. We recognize the long dorsal fin, springing crest-wise at the front, running the length of the body like an extended Mohawk hairstyle.

Mahimahi's modern name might suggest an endemic link to Hawaii (*mahi* is Hawaiian for "strong"; duplicating the word intensifies the meaning). The fish is in fact present in many tropical, subtropical and temperate zones. It's been found as far north as the Sea of Okhotsk. Somewhat confusingly, mahimahi is also known as dolphinfish; there's a resemblance about the forehead, though any biological commonality with that mammal is, of course, nil.

A mahimahi's life is spent in fast-forward mode. It grows extremely fast, maturing sexually within four months or so. It also dies young, at two or three years of age. In between, it can reach 2 metres in length – a powerful, nervy wallop of a fish, capable of acrobatic resistance when caught. Mahimahi is also photo-mutant. Thanks to cells known as chromatophores, its skin changes colour in response to external stimuli. When out of water, the fish glows a psychedelic mix of yellows, blues and greens, specked with ultramarine. As it skims the waves, you might mistake it for a surfboard. Postcapture, when the nervous system shuts down, the colours fade.

Worldwide presence aside, mahimahi is mostly consumed in the United States of America and the Caribbean, though it's not uncommon in Japan. This is a prized fish, its texture flaky but firm, its flavour characterful yet light – reminiscent of swordfish, but free from its oily pungency. True, tasty though it is, mahimahi is not ravishingly unique on the palate, nor is it essential to global food security: a case might be made that such a robust charismatic creature is better admired – or, indeed, painted on villa walls – than eaten. Then again, even if left alone in the wild, mahimahi scores poorly for longevity. And fishing it does provide essential income for some Caribbean beachside communities – for example, in parts of the Dominican Republic, where our recipe comes from.

Blue, yellow and short-lived

KNOW YOUR FISH

Mahimahi will almost always be sold as fillets, skin-on or skinless. In the United States of America, availability will be highest in the early months of the year, fed by strong exports from Ecuador and Peru. This seasonality means that prices may vary significantly. Look for flesh that is a subdued pink, occasionally with a sharply delineated, crimson bloodline. Darker meat always correlates with stronger taste, so you may wish to cut out those redder portions. The skin is tough and leathery: it's generally removed before cooking.

Nutrition facts

MAHIMAHI, FILLET, RAW
per 100 grams

ENERGY (kcal)	100
PROTEIN (g)	22.1
CALCIUM (Ca) (Mg)	9
IRON (Fe) (Mg)	0.5
ZINC (Zn) (Mg)	0.4
VITAMIN A (RETINOL) (μg)	4
VITAMIN B12 (μg)	0.6

My, how you've grown.
I remember you the size of
my thumb....
Well, it was a very long time ago.
Four months or more!

[HIDES SMILE] **You look like
a glorious technicolour
dolphin.**
I don't go in for that grey dolphin
shade much. Also, I'm definitely a
fish, not a mammal: if I look like a
dolphin, it's just coincidence. You
know how sometimes an apple can
look like a pear.

*But the apple and the pear
are both fruit.*
Okay. Ignore the analogy. You get
my drift. In any case, even if you
strip away the colour, the dolphin
has that long beak...

*Yes. It's technically called
a rostrum.*
... and none of this bristly fin that
you see rising from the top of my
head and running down to my tail.

*I've been meaning to ask you
about that. It does give you
a warrior look.*
That's just as well, because I am
one. You have to be. It's dog eat dog
down here.

*I think you mean that the
big fish, like you, eat the
small fish. You keep mixing
your analogies.*
Well, how articulate were you at
my age?

*Fair enough. Will I be seeing
you tonight at dinner?*
Not if I have anything to do with it.

THE INTERVIEW
MAHIMAHI

Samaná–style coconut fish

PESCADO CON COCO ESTILE SAMANÁ

The Dominican Republic hauled in nearly 500 tonnes of mahimahi *(dorado)* in 2018, a small proportion of which was shipped to the United States of America. The entire catch, that said, brought in a relatively modest USD 700 000. Starting in 2021, FAO began working with fisherfolk in the south of the country to improve the cold chain, open up new markets and plug artisanal mahimahi fisheries into the tourist industry.

Dominicanos still eat relatively little fish, but do so with tropical verve. Samaná, on the country's northeastern coast, is an enticingly lush peninsula of wild beaches, rainforests and coconut plantations. The area, in fact, claims to feature the highest concentration of coconut palms in the world – and the local cuisine seems to bear out that metric. Samaná-style coconut fish, the area's trademark recipe, is not a complex dish. But it shines with the unpretentious perfection of a lazy lunch at the water's edge. You may add, if you wish, *achiote*, the Caribbean condiment that gives the sauce both an attractive red colour and a spicy, telluric warmth – or, failing this, smoked paprika.

If mahimahi is unavailable, swordfish or grouper, or another firm, characterful white fish will do. Accompany your *pescado* with cold lager or – if you prefer wine – a chalky dry white.

TO FEED
4 PEOPLE
YOU WILL NEED

4 FILLETS MAHIMAHI OR OTHER FIRM WHITE FISH

1 RED ONION, JULIENNED

1 CAPSICUM, JULIENNED

2 TOMATOES, PEELED AND CUBED

(You can easily peel them if you leave them in boiled water for a few minutes.)

2 CLOVES GARLIC, FINELY CHOPPED

OLIVE OIL

I PINCH OREGANO

30 ML (2 TBSP) TOMATO PASTE

450–500 ML COCONUT MILK
(the cooking variety – about one can, plus a quarter of another)
•
250 G FLOUR
•
SALT AND PEPPER
•
OLIVE OIL FOR COOKING

METHOD

1 Season the fish fillets and dust them in the flour, then seal them quickly with oil in a frying pan. Drain them and set them aside.

2 Into the same oil, pour the onion, capsicum, garlic, tomatoes and tomato paste and simmer on a medium flame to form a fragrant base.

3 Now pour in the coconut milk, turn up the flame and bring to a low boil. Turn down the flame again and simmer until the sauce has reduced by half. Season as needed.

4 Place the fish fillets in the sauce and cook them gently for 2–3 minutes per side. Serve with rice or *mangú*, the Dominican plantain mash.

Pomfret

BRAMA BRAMA, PAMPUS ARGENTEUS

Pomfret (Bramidae family) fall into some 20 species – though the common name may further include, or overlap with, an array of closely related butterfish. Distributed around the world's oceans, and as far north as Norwegian waters, these are highly migratory fish that tend to favour small schools. They're flat in shape and light silver to black in colour. Seen in profile, a pomfret's body is a near-perfect oval, which an aerodynamic set of fins may turn, in some species at least, into a near-perfect lozenge. This rather appealing geometry is spoiled somewhat by the mouth, a deep gash into the front of the animal that gives it a half-perplexed, half-irate expression.

Widely distributed though they are, the popularity of pomfret varies greatly with the region. In Europe, the fish is little known, despite a fair presence in the Mediterranean. In FAO's Italian homeland, supplies are erratic at best, and the *pesce castagna*, or "chestnut fish", absent from restaurant menus. One local fish inspector testifies to the mix of obscurity and meagre esteem afflicting the fish. Even when available, the colour of pomfret (which tends to darken post-capture) puts people off, he says. Retail prices are accordingly modest. Unjustifiably so, our inspector adds: "Do get the *pesce castagna* if you find it fresh. Those fillets will land you with a couple of juicy, high-quality fish steaks."

Pomfret flesh is indeed delightfully textured, semi-springy and sweet, on top of being low in calories and rich in iron and phosphorus. It grills spectacularly. And if Italians and other Europeans are largely resistant to pomfret's appeal, the fish is much enjoyed in South Asia, and along Atlantic and Pacific seaboards. Reassuringly, pomfret is listed as "of least concern," the lowest-risk category – though this status conceals uneven set-ups. The Atlantic variety

A *fish* unjustly ignored

(*Brama Brama*) is generally plentiful. By contrast, the silver pomfret (*Pampus argenteus*) is in high demand on the subcontinent under the name of butterfish, so is becoming scarcer and dearer. The *People's Archive of Rural India*, an online journal and archive, has movingly documented the shrinking of the pomfret around Mumbai and its social fallout: here, a combination of pollution, mangrove clearance and overfishing is forcing the fish ever further from shore. Meanwhile, global warming is speeding up pomfret's biological clock, causing early maturity and reduced size.

So eat pomfret, by all means: it would be a shame not to. But if you're conservation-minded, you might wish to up your intake in those parts of the world where pomfret is unjustly overlooked, and moderate it where the fish could do with a break.

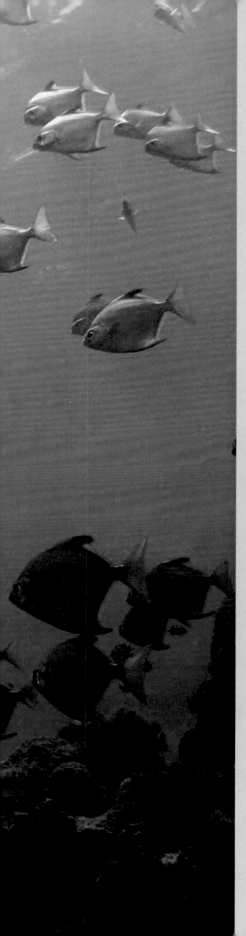

KNOW
YOUR FISH

Pomfret offered for sale will rarely exceed 40 centimetres (up to a rare maximum of 60), and may sometimes be found at half that size. Because of these dimensions, it's generally sold whole, perfect for grilling or baking.

If you fancy barbecuing your pomfret using a long-handled fish rack, you could snip off the fins and tail, then sever the head in a straight line, vertically across the fish, just past the eye. Indian cooks will sometimes slide the blade transversally under the eye and towards the mouth on either side, removing the head in the form of two orbital slices. Alternatively, when preparing a "curry cut," they may keep the head, with just the eyeballs taken off.

When this is done, get rid of the crimson gills, then make deep parallel incisions down the side of the fish, running from back to belly, so that your pomfret looks like a perforated pouch: this will allow you to slide a couple of fingers in and pull out (and discard) the stomach lining. Wash the fish, then smear it inside and out with a tandoori sauce of yoghurt mixed with salt, shredded ginger, crushed garlic, chilli, garam masala, lemon juice and powdered turmeric before you place it on the barbecue: the pomfret should come out enticingly charred, red-orange streaked with black.

Or follow our featured recipe, which uses the fillets of the fish in a wetter, wine-spiked, all-in seafood extravanganza.

Nutrition facts

FISH; SILVER POMFRET; MUSCLE TISSUE, RAW
per 100 grams

ENERGY (kcal)	175
PROTEIN (g)	17.1
CALCIUM (Ca) (Mg)	21
IRON (Fe) (Mg)	0.3
ZINC (Zn) (Mg)	0.5
SELENIUM (Se) (μg)	0
VITAMIN A (RETINOL) (μg)	90
VITAMIN D3 (μg)	5
VITAMIN B12 (μg)	1.4
OMEGA-3 PUFAS (g)	1.23
EPA (g)	0.24
DHA (g)	0.65

Hello, my rhomboid friend.
I wasn't sure what you wanted from me. Why am I here?

I just wanted to know you better.
But why? I'm small. I'm not particularly handsome. They don't like me much in Europe: they've even said I look like a chestnut. I only tend to appear at the market when I turn up as bycatch. It's not that nice, being an "incidental" fish, you know.

I don't think being likened to a chestnut is really that derogatory. But I do agree that you deserve to be better known in parts of the world.
It's kind of you to say so. You're writing a book, I hear? Give me a good write-up.

I have.
Did you mention *monchong*?

No, what is that?
It's what they call me in Hawaii. They like to serve me in tacos, or with a papaya salad. And you know, they like pineapple with things there. So I've wound up in that scenario too.

Thanks, I'll make a note of it. In the course of my research, I also discovered that there's a town that bears your name: Pomfret, Connecticut. Quite pretty, apparently. Population 4 266, according to the most recent US census.
Oh? Well, there are worse ways to go down in history. A marine greeting to any readers there!

THE INTERVIEW
POMFRET

Pomfret in margarita sauce

REINETA EN SALSA MARGARITA

Despite having one of the world's longest coastlines (the country is, in effect, one extended coastline), Chileans eat relatively little fish – well below the world average at 12 kilograms a head or so, and certainly much less than they do beef and other meats. Most of what Chile catches, it sends abroad. Salmon alone makes up almost half of the nation's food exports and takes second place of all exports behind copper.

For all of its external popularity, however, salmon faces strong competition in its home market from hake and pomfret. And, as if to belie the paucity of national seafood consumption, this recipe piles on the species: joining the pomfret are clams, mussels, oysters and prawns. (Some of the shellfish may be omitted, but dropping all of it would ruin the layered delight of the dish. The pomfret, on the other hand, can be replaced with other oceanic white fish.) *Salsa margarita*, a Chilean speciality, combines the seafood juices with flour, butter and milk – creating, in effect, a marine béchamel. Note that in Chile, crustaceans and bivalves are often sold shelled and prepacked, and even precooked.

TO FEED
4 PEOPLE
YOU WILL NEED

4 POMFRET FILLETS

200 G MUSSELS, SHELLED

200 G PRAWNS,
FULLY SHELLED, HEADS OFF

100 ML BUTTER

240 ML MILK

I ONION,
SLICED LENGTHWISE
FROM STEM TO ROOT

20 G FLOUR

200 G OYSTERS, SHELLED
•
120 ML DRY WHITE WINE

METHOD

1   Preheat the oven to 180° C. Place the fillets in an oven dish and cover them with the wine. Season with salt and pepper. Spread the sliced onion around the dish, and also most of the butter in chunks. Bake for 12–15 minutes.

2   Place the remaining butter in a saucepan, adding the milk and sifted flour. Over a medium flame, whisk into a sauce. Add the shellfish, with a little of the cooking juices from the fish fillets. Season and simmer for 2–3 minutes.

3   Remove the fish from the oven dish, place it on a serving dish, and cover it with the white sauce and the shellfish.

Salmon

SALMO SALAR

If Aquaculture were a job applicant, her résumé would list salmon as a key career achievement. If she'd had today's economic heft 150 years ago, she might have been represented allegorically on the façade of civic buildings, cradling a salmon in her arms. Judging by sheer quantity if nothing else, *Salmo salar* – Atlantic salmon – is the crowning success of fish farming. It is its vindication, its bounty. We see farmed salmon at every turn: in a fish pie or sandwich or bagel; on delivery pizza; shredded through pasta; as tartare, or as sashimi, or in a California roll.

Faced with this omnipresence, we may want to rewind a little and remind ourselves where this fish, so attractively pink-fleshed as to have a colour named after it, comes from. Its distribution in the wild mirrors almost exactly that of cod – before cod thinned out. From Arctic Russia, salmon grounds swoop along western Europe's Atlantic seaboard and into the Baltic, fan out to Iceland and Greenland, embrace the Grand Banks and Maritime Canada, and come to hug the coast of New England. To which we must add – totally unlike cod, this – the river system of Europe and North America. Salmon are

anadromous (from the Latinized Greek for "upward-running"): they hatch in freshwater, migrate to the ocean, and return upriver to spawn. As they change habitat, their livery veers between a sea-compatible silver and a more riverine camouflage tinge. After spawning, they die, though not always.

Atlantic salmon has relatives, some "true" salmon, some not. They include Pacific species such as the sockeye (*Oncorhynchus nerka*), whose body (but not head) turns a deep cyclamen during spawning. Some varieties are lakelocked. Some shade off into trout, a close cousin: within the broad Salmonidae family, biological boundaries can be fluid. But of *Salmo*, there's only one. This is the one we know, in common parlance, as "salmon," with no qualifier needed. This is the one we enjoy caught in the wild, with its deep pink, quasi-orange shade from feeding on red krill; and more often farmed, with its lighter, more marbled flesh.

Farmed to rule

The extent to which salmon has been democratized, and the speed of it, has few parallels in history. Global consumption has tripled in one generation, and most of it is farmed fish. In 2019, the BBC reported that UK residents consumed one million salmon meals every day. This, in a country where wild salmon is no longer fished commercially. Salmon aquaculture, in fact, has been described as the world's fastest-growing food-production system. It's as if we saw something in nature, something luxurious and desirable, and 3D-printed it to our heart's content. In the process, salmon has been put to entirely new uses. Thirty years ago, the notion of salmon sashimi was still incongruous to most Japanese. The species was thought too fatty, prone to spoilage and parasites – until Norwegian salmon farmers, looking to mop up overproduction, conjured it as a marketing strategy. This wasn't happenstance: Norway is credited with having the highest-quality strains of wild salmon, and hence the best fingerlings for hatcheries. The country is by a long stretch the world's largest producer. Next up is Chile. Between them, the two nations cover two-thirds of the market.

Salmon farming has had its share of bad rap, largely to do with the low mobility of the fish in pens and cages; with the occasional escape of farmed animals, which can theoretically weaken wild populations; or with contamination of the sea, where insufficient water flows cause concentrations of salmon excrement.

As in other cases involving aquaculture, the criticism is part valid, part not, and part dated. Excessive crowding

does happen. But the idea of having the fish still or barely hovering, like straphangers in the commuter crush, runs counter to market logic. Salmon is a capital-intensive, mass premium product: farmers have every interest to grow the fish to the ideal market size in the most efficient way. This entails keeping them healthy. Conversely, packing cages too tightly is counterproductive: it means shelling out more for costly fingerlings and reaping lower profits from smaller, weaker animals.

Still, farmed salmon are by definition more static than wild counterparts. In Norse lore, salmon was a symbol of sprightly elusiveness: the prankster god Loki, a master shapeshifter, is said to have turned himself into one to escape other gods' wrath. Today's farmed salmon are fatter and dowdier. Then again, they tend to be more nutritious: thanks to their optimized diets, unavailable in Viking times, they can be richer in omega-3 acids. (Farmed salmon are fed something called kibble, a combination of vegetable meals and fish-based meal and oil.)

It's not unheard of for farmed animals to escape. That said, salmon pens and cages tend to be placed in bays, fjords or lochs, where outer islands protect them from oceanic storms and winter gales. Submergible cages, more resilient to disruption from waves, have been developed. The very severity of escape incidents is meanwhile being questioned. In 2022, Chile's Constitutional Tribunal threw out heavy fines levied against one island farm which had seen more than 600 000 salmon break out of their pens during a violent storm. The judges ruled there was no scientific proof of environmental damage in escape cases, and that presuming detrimental impacts was unconstitutional.

As for sea contamination from large quantities of excrement, farming sites can be rotated – much like in agriculture, where fields are left fallow to allow soils to regenerate. One newer, more pristine option is to sever the marine link entirely. Research money has poured into land-based salmon farming, a closed-loop technology based on recirculation aquaculture systems (RAS). This allows for almost completely controlled environments. In 2020/21, RAS salmon firms were a sensation on the Oslo stock exchange. But with kinks still to be ironed out – the process depends on constant, finely calibrated oxygen and electricity supplies: an accident at one farm in Denmark, in the summer of 2021, wiped out almost a fifth of its biomass – investors have grown warier and shares have fallen back.

Facing down the critics

KNOW
YOUR FISH

Salmon can grow to 1.5 metres, which means you'll rarely come across one at the fishmonger's, let alone in supermarkets. The standard offer will involve fillets, fresh and thickly cut for steaks, or thinly sliced and prepacked if the fish is smoked. In the United Kingdom, it's not uncommon to find cubed salmon as part of a fish-pie mix, alongside chunks of haddock and cod. Ready-meal sushi boxes containing salmon rolls, nigiri or sashimi have also become ubiquitous in western markets.

Smoked salmon, whether wild or farmed, will feature the country of origin on the packet. The wood used in the smoking will also likely be mentioned – an expanding choice from alder to maple or juniper. Lox, while often thought synonymous with smoked salmon, is a distinct salt cure closely associated with New York's Jewish community: the fish may be smoked, but not necessarily.

Farmed salmon would normally be a shade of beige. As in the case of trout, the pink comes from astaxanthin, a red pigment added to the feed. This is a safe, authorized compound, and while in the case of cultured fish it would be synthesized, it's also what gives wild salmon its natural colour. If the shade is much deeper in wild animals, it's because they move much more, and are thus more efficient at converting feed to flesh.

Farmed salmon is rosier and, with its higher fat content, often streakier in appearance. Fish with a lot of white showing through may taste greasy; the flavour will, in all circumstances, be less gamey than that of captured fish. In both wild and farmed salmon, the grey-brown flesh under the skin has an intense tang: many producers and retailers remove it along with the skin, although it is, of course, perfectly innocuous health-wise.

Speaking of health, antimicrobial resistance in humans is a growing concern. The world's leading fish farm certification scheme, the Aquaculture Stewardship Council (ASC), takes a hard line on antibiotics being administered preventatively, or on their use to promote growth or for other medically unnecessary ends. Through vaccination and non-pharmaceutical methods such as ultrasounds to combat sea lice, which are salmon's most threatening parasite, Norway has largely succeeded in eliminating antibiotics from the food chain: records show that in 2020, when the country produced a million tonnes of farmed salmon, fewer than 50 prescriptions were issued. The next-largest producer, Chile, remains more partial to antibiotics – but there too, their prevalence is diminishing.

Nutrition facts

ATLANTIC SALMON, WILD, FILLET W/O SKIN, RAW (N.S.)
per 100 grams

ENERGY (kcal)	177
PROTEIN (g)	20.1
CALCIUM (Ca) (Mg)	15.7
IRON (Fe) (Mg)	0.7
ZINC (Zn) (Mg)	0.6
IODINE (I) (μg)	21
SELENIUM (Se) (μg)	30
VITAMIN A (RETINOL) (μg)	16
VITAMIN D3 (μg)	14
VITAMIN B12 (μg)	5.3
OMEGA-3 PUFAS (g)	2.64
EPA (g)	0.55
DHA (g)	1.71

ATLANTIC SALMON, FARMED, FILLET W/O SKIN, RAW (N.S.)
per 100 grams

ENERGY (kcal)	201
PROTEIN (g)	19.9
CALCIUM (Ca) (Mg)	11.8
IRON (Fe) (Mg)	0.3
ZINC (Zn) (Mg)	0.4
IODINE (I) (μg)	9
SELENIUM (Se) (μg)	21
VITAMIN A (RETINOL) (μg)	9
VITAMIN D3 (μg)	7
VITAMIN B12 (μg)	4.1
OMEGA-3 PUFAS (g)	2.59
EPA (g)	0.74
DHA (g)	1.25

Hi. Are you wild or farmed?
I do wish the world would stop creating these divisions. Driving these wedges, you know.... We're all people!

But you're not. You're fish!
I know. It's a manner of speaking. The argument stands.

You mean to say there's no difference between the wild and farmed versions of you?
Of course there is. And it can be relevant in some contexts. But I'm speaking as a general gastronomic proposition. The farmed ones among us have taken a lot of flak. Faced a lot of prejudice, often unjustifiably.

Point taken. If we look beyond the divide – how have you dealt with your surge in popularity around the world, in the last three decades or so? Has that taken a big toll on you?
Listen, much as I'd like to play the damaged star, it's been great. It's wonderful to be liked, to reach new audiences. I must confess I was a bit surprised to turn up as sashimi, some years ago. Wasn't sure I'd cut it. But it's worked out well.

Is there anything you resent about the way you've been treated?
Being overcooked. Still happens a lot. Terribly unpleasant. Brings out the worst in me.

I'm sure. We've given clear instructions in that regard in our write-up.
Such a good sport. I knew I could count on you.

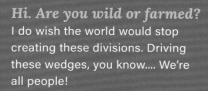

THE INTERVIEW
SALMON

Lemon–baked salmon with vegetables and boiled potatoes

CITRONBAKAD LAX MED GRÖNSAKER OCH KOKT POTATIS

Despite dampened investor enthusiasm for the technology, funding was secured in 2021 to build Sweden's first land-based salmon farm. Currently, all farmed salmon eaten in Sweden is imported from neighbouring Norway – though Swedes also consume some salmon from capture. The fish appears in many guises: poached or sugar-cured, in dill and mustard sauces, or as *laxpudding* – in a gratin, crusty and golden on top, soft and creamy below.

Our recipe features an example of *husmanskost*, or traditional home-cooking. This is a fish-spuds-greens-and-dairy combo, a simple enough dish, but one where the skill consists (as it often does) of not overcooking either salmon or vegetables. Until you get the hang of it, it's best to ensure that you have a food thermometer. The internal temperature of the fish should not go above 45° C – or else the albumin, a protein found in liquid form inside the flesh, will congeal. This accounts for the blobs of white gunk on the surface of cooked salmon which, if your experience of office canteens and other mass catering mirrors ours, will have sent you scrambling for a different lunch option. If you'd rather not fiddle with a thermometer, take a leaf from the book of lox and brine your salmon before cooking: the process helps concentrate flavour and seal in moisture.

TO FEED
4 PEOPLE
YOU WILL NEED

600 G POTATOES

200 G BROCCOLI

4 SALMON STEAKS

200 G SUGAR PEAS

250 G BUTTER

2 ONIONS, CHOPPED

1 LEMON

100 ML CREAM

SALT

100 ML FISH STOCK

100 ML DRY WHITE WINE

METHOD

1 Melt 50 g of butter over a soft flame and mix with half the lemon juice, then rub the mixture all over the salmon. Place the fish in an oven dish and bake at 100° C, to an internal temperature of 45° C. While the fish is cooking, boil the potatoes.

2 On the hob, fry the chopped onions in a little butter until they are soft, then add the wine, fish broth and cream, then the remaining lemon juice. Bring to a simmer and reduce to a generous sauce consistency. Strain and keep warm.

3 Chop the broccoli into small pieces and place in a saucepan with the sugar peas, add 50 g of butter and a good pinch of salt. Soften the vegetables for a few minutes, but make sure the broccoli keeps its crunch.

4 Now add the remaining butter, cold, to the warm sauce and whisk until frothy. Plate up the fish and vegetables with the boiled potatoes, spooning the sauce on top.

Seabass

DICENTRARCHUS LABRAX

....

Seabass may well be globalization's poster fish, a signifier of mass exclusivity from Athens to Vancouver, like Belgian chocolates or an *Economist* subscription. It certainly tastes very good. The flesh structured and buttery, its marine flavour distinct but politely reined in, you could describe seabass as a larger, preppier, upwardly mobile sardine. This, of course, is only a functional analogy: biologically, seabass is nowhere near a sardine.

Then again, seabass is not really one thing. Nearly 500 species of the vast Serranidae and Moronidae families may be crammed under that label. Overall, they tend to be elongated ocean fish, silver to grey or even black in colour. But there is also much variance. You may find seabass just a few centimetres long, or a stonking two metres on occasion, or anything inbetween. The category will unambiguously include the European seabass (*Dicentrarchus labrax*), which hugs the Atlantic shores from southern Sweden to Senegal, and extends through the Mediterranean into the Black Sea. Yet "bass" may also cover a

clutch of American species – such as the white seabass (*Atractoscion nobilis*, also known as weakfish) – which are actually groupers or croakers (for a discussion of grouper, see p. 72); or the Asian barramundi (*Lates calcarifer*), wildly popular in Australia, often in the guise of fish-and-chips; or else a big American freshwater species, the *Morone chrysops*, frequently hybridized with a marine species to create "striped bass"; and many more.

Of all claimants to the name, the European *labrax* is the one most commonly labelled seabass for commercial purposes. This is an intensively farmed fish, cultured for the most part in net-pens in the eastern Mediterranean. Türkiye and Greece supply two-thirds of the market, worth around a quarter of a million tonnes a year, with Egypt some way behind. At the Mediterranean's opposite end, Morocco and Portugal also produce it.

Seabass tends to be sold at 30–50 centimetres long, just the right size for serving whole on a plate or a table dish. It's an immensely versatile food fish, good fried or braised; roast or poached; or done in a *court bouillon* of diced carrot, celery and onion; or else raw as sushi and sashimi. Its flesh – low in calories, high in protein and rich in vitamin B6 – is oily enough to keep it moist even when (slightly) overcooked.

The first-century author Pliny the Elder, in his sometimes fanciful *Natural History*, tells us that "violent animosity rages between the mullet and the seabass," but there's otherwise little known lore attached to the *labrax*. It's also the case that outside the Mediterranean, the fish is a relatively recent discovery: it doesn't, so to speak, have a storied personality. What it does have, like all fresh celebrities, is imitators. In 1977, in a move that became a case study in marketing strategy, an entirely unrelated animal, the Patagonian toothfish (*Dissostichus eleginoides*), was ingeniously rebranded as "Chilean seabass" for the benefit of US consumers. This fish was in reality a type of cod; and while succulent, it was exceptionally ugly to look at. Yet in a market dominated by fish fillets, the aspect of the Patagonian toothfish, with its utter dissemblance to seabass, had no impact. Thriving on the ambiguous nomenclature, the fish became so popular as to be brought to the edge of extinction. A restaurant-led backlash has since reduced pressure and allowed stocks to rebuild. In the process, management techniques have improved, labelling rules have tightened, and consumer awareness has shot up. "Chilean seabass," that convincing interloper, is sustainably back on world menus.

A *polite bass line*

KNOW
YOUR FISH

Fetching as it does fairly high prices, seabass is thought to be one of the most frequently mislabelled fish.

A 2019 investigation by Oceana, an organization campaigning against seafood fraud, put seabass at the top of the list, with a 55-percent mislabelling rate across grocery stores, markets and restaurants in the United States of America. (The rate for Canada, published a year before, was 50 percent – though here seabass took second place to red snapper.) Commenting on the findings, Oceana said the substitution of lower-value fish, such as tilapia or Asian catfish, for higher-value ones could "mask health and conservation risks".

The numbers, though, have not gone unchallenged. Critics took issue with what they saw as selective sampling; with the use of allegedly outdated DNA reference bases; or with error being passed off as intent, in a sector beset by language barriers and multiple accepted market names. It's certainly true that with a variety of species qualifying to some extent for the "bass" moniker, the space for labelling overlap is great. No one will deny that being deliberately sold cheap tilapia in lieu of seabass is indefensible. By contrast, getting, say, a fine grouper by mistake is much less of a concern. In Italy, where our recipe comes from, the *spigola* (seabass, also known as *branzino*), the *orata* (seabream) and the *ombrina* (croaker) are gastronomically interchangeable for the most part. All are quality fish: they may vary a little in looks – the bream and croaker more compressed in shape than the seabass – but they rival each other for subtlety and succulence. Thanks to aquaculture, all three are available year-round.

Nutrition facts

FISH, SEA BASS, MIXED SPECIES, RAW
per 100 grams

ENERGY (kcal)	97
PROTEIN (g)	18.4
CALCIUM (Ca) (Mg)	10
IRON (Fe) (Mg)	0.3
ZINC (Zn) (Mg)	0.4
SELENIUM (Se) (μg)	37
VITAMIN A (RETINOL) (μg)	46
VITAMIN D3 (μg)	6
VITAMIN B12 (μg)	0.3
EPA (g)	0.161
DHA (g)	0.434

How should we go about this, dear seabass? Where should we start?
I don't know. Is it harder than with other fish? Is there something wrong with me?

No, no, not at all. You taste great, but you also seem a little... insubstantial. Forgive me for saying so. You're a fine fish, yet it's not easy to find a salient fact about you. You don't make a biographer's task easy.
You may be right. I've been told I'm very well-bred but a little lacking in character. I can't really point to singular achievements. Never acted up, always got good marks, didn't cause a stir. But maybe thanks to that, I end up on many fine or upwardly mobile tables. I see it as a good measure of success.

Certainly – hence your presence in this book. You also fan out into a very large number of species. Would you say that makes for a rather de-centred personality?
I can't say. I'm not really well up on critical theory and that sort of thing.

You imply you never got into trouble, but Pliny does have this rather cryptic line about your supposed violent animosity with mullet. What's that about?
Oh, nothing, really.

I must insist.
Well, there *was* a bit of rivalry about appearing in a seafood mosaic at Pompeii, now at the National Archaeological Museum in Naples. It went to arbitration: in the event, we both got a spot. Go have a look. You could be there in two hours from FAO headquarters on the high-speed train.

THE INTERVIEW
SEABASS

Seabass in a potato and courgette crust

SPIGOLA IN CROSTA DI PATATE E ZUCCHINE

One classic way to prepare *spigola* **in Italy is** *all'acqua pazza* **(with "crazy water").** Paternity for this dish, which used seawater back when salt was a pricey ingredient, is disputed between the fisherfolk of Naples and the islanders of Ponza. It involves baking the seabass in salted water and wine, with cherry tomatoes, crushed garlic and parsley. The cooked fish is brought whole to the table and shredded onto individual plates, with a ladleful of the savoury broth poured on top.

At the more *recherché* end, *La Cucina italiana*, Italy's culinary magazine of record, suggests a seabass raspberry confit. To serve four, heat up peanut or another neutral cooking oil to exactly 60 degrees in a saucepan, then turn off the flame. Cut four seabass fillets into three slices each. Salt the pieces of fish and place them in the warm oil for 20 minutes, ensuring they're completely covered. Separately, whizz 100 grams of raspberries in a blender with four tablespoons of olive oil, two tablespoons of

apple vinegar and a pinch of salt. Remove the seabass fillets from the oil, drain them on kitchen paper, and dress them with mixed leaves and piped drops of raspberry sauce. Toss more fresh raspberries over the dish, and serve with a grind of fresh pepper.

Our own recipe keeps things a little more rustic. The fact is, outside of more adventurous restaurants, a side of roast potatoes is rarely absent when you order fish in Italy. This iteration places the spuds under the fish and the courgettes on top, a burst of comfort food (admittedly with a playful touch) enlivened by earthy oregano. As the courgette will

obscure the fish, make sure your fillets are thoroughly deboned: use tweezers if necessary. Jazz up the finished dish with a sprinkle of bottarga, the dried fish roe condiment, if you have any handy.

Since our recipe provides both starch and vegetable content, no accompaniment is needed, strictly speaking. But you might want to look in the direction of bitter leaves. A bowl of chicory, sautéed with chilli and garlic, works well. The classic Italian tipple to have with *spigola* is a light Ligurian Vermentino. Or you could offset the simplicity of what is, in effect, a fish pie with a classy glass of fizz from Trentino or Piedmont.

TO FEED
4 PEOPLE
YOU WILL NEED

4 SEABASS FILLETS

1 LARGE POTATO, PEELED

**2 COURGETTES
(ZUCCHINI)**

150 G BREADCRUMBS

OLIVE OIL

DRIED OREGANO

SALT AND PEPPER

METHOD

1 Preheat the oven to 180° C. Separately, grease an oven tray (or large glass ovenproof dish).

2 Slice the potato thin and arrange the slices in a broad rectangular pattern in the oven tray. Make sure that the slices overlap a little to provide continuity. Season them and brush them with olive oil.

3 Now place the fish fillets over the potatoes, next to each other so that they touch, two down and two across. Cut and adjust the edges of the potato layer to ensure the potato rectangle matches the rectangle of fish that sits on top of it. Season the fish.

4 Next, grate the courgettes on top of the fish, ensuring again that the shape matches that of the fish and potatoes. You should end up with a large, flat sandwich, of which the seabass is the filling. Tap them down to maximize density, but don't squeeze the courgettes too much: their water content will ensure the fish stays moist.

5 Sprinkle the breadcrumbs all over the courgettes, with oregano and a drizzle of olive oil. Roast in the oven for 20 minutes or so, then take out of the oven and let rest for 2 minutes. Cut into four and serve.

Snapper

LUTJANUS CAMPECHANUS

Brow arched, mouth sloping glumly at the edges, snappers (Lutijanidae) seem to go through their long life – as long as 80 years, new research suggests – in a state of reproachful irritation. This impressive life expectancy is, of course, largely theoretical: carnivorous and sharp-toothed as they may be, snappers are some way off the top of the food chain. They are much liked, in fact, by sharks, barracudas and humans.

Of the over 100 species distributed around the world's tropical and subtropical areas, the most sought-after is the red snapper (*Lutjanus campechanus*). The fish is at its densest in the reef-rich waters of the western Atlantic, chiefly in the Gulf of Mexico.

Red snapper is both succulent and, in its bright red livery, supremely photogenic. The flesh is firm, protein-rich, and low on sodium and saturated fat. The flavour – mellow, vaguely nutty, barely piscine – has quasi-universal appeal, including with children and fish-sceptics. Some US restaurant menus end up offering red snapper at USD 40 per fillet, in turn creating an incentive for poaching in neighbouring countries. But while overfishing is rife in some areas, vigorous regulation on the Gulf's northern rim has helped populations rebound since the early 2000s. At the time of writing, the US commercial season is limited to six months a year, and the recreational season to just three days.

There's more good news: in 2021, a multiyear study commissioned by the US Congress tripled previous estimates of red snapper abundance. Led by the Harte Research Institute, a body that monitors the Gulf of Mexico's ecosystem, the "Great Red Snapper Count" involved a new mix of tagging, direct visual counts, habitat classification, advanced camera work and hydroacoustic surveys: it concluded there were well over 100 million red snappers, vast numbers of them loafing happily around the Gulf's unmapped bottom habitats. (Earlier numbers had been extrapolated from biomass estimates.)

Red snappers will easily grow to 1 metre long – but are also commonly (and legally) fished in early adulthood, coming in at 38–40 centimetres. This means, rather gratifyingly, that a smaller exemplar will fit inside a family oven, while a larger one may feed a banquet.

And now, to cooking. Red snapper catches the eye, so it pays to leave the skin on. For a dramatic effect, take a black slate platter and dust it with ground pistachio. Separately, put a courgette through a spiralizer or meat grinder. Dress the resulting courgette vermicelli with olive oil, then season it with salt, pepper, a quarter teaspoon of sweet paprika and a pinch of cinnamon. Place it on cling film in a thick sausage shape, wrap the cling film around it and roll it further into a smooth tight cylinder. Freeze this courgette cylinder for 15 minutes or so to harden it, then take it out, strip off the cling film and cut a 5-cm-long section, ensuring plane surfaces at both ends. Stand the courgette cylinder at one end of your black platter.

Move on to the fish: grill a red snapper fillet, scoring and crisping the skin but keeping the fish moist below. Stack the fillet carefully on top of the courgette cylinder, skin side up. For a final touch, soak a couple of whole pistachio nuts in lemon juice and dry-fry them for a few seconds, then place them on top of the fish.

Or give yourself a break from the visual acrobatics and follow one of the homelier, but equally satisfying recipes in this chapter. Either way, make sure you read the next section to avoid getting scammed: red snapper is among the most faked and mislabelled seafood on the planet.

Painting the ocean red

KNOW
YOUR FISH

Depending on surveys and sampling locations, a third to near-100 percent of what is sold as red snapper may not be that. The highest mislabelling rates have been associated with sushi restaurants, where the fish on your plate is least identifiable. This needn't cause a headache if what you're getting is a closely related, hardly distinguishable snapper species, of which there are a fair number – crimson, rose, vermillion or yellowtail. (One of the two recipes listed here originates from Baja California, home to the Pacific snapper: this species sports slightly different genetics despite similar chromatics.) But mislabelling does become a major problem – and potentially a criminal one – when what you're getting is, say, tilapia, a farmed, mostly freshwater fish of far lesser commercial value.

To minimize the risk of snapper swap, don't buy anything that hasn't got its skin on. This really is a case of judging a fish by its cover. True, there's a definite blush to the flesh of red snapper too, but not enough of it for a clear call. The proof lies squarely on the outside: the skin should at the very least be a distinct pink, and often (especially along the dorsal fin) a pomegranate-hued blaze.

If buying the fish whole, also inspect the eye. It too should be bright red. Red too, the tailfin, which is moreover slightly forked – unlike tilapia's greyish, brush-like one. Finally, should you come across red snapper enthusiastically advertised as "fresh off the dock," pull out your phone and run a search for the local legal fishing season.

None of this is fail-safe, of course. But a minimum of due diligence helps.

Nutrition facts

SNAPPER, MIXED SPECIES, RAW *per 100 grams*	
ENERGY (kcal)	100
PROTEIN (g)	20.5
CALCIUM (Ca) (Mg)	32
IRON (Fe) (Mg)	0.2
ZINC (Zn) (Mg)	0.4
SELENIUM (Se) (µg)	38
VITAMIN A (RETINOL) (µg)	32
VITAMIN D3 (µg)	10
VITAMIN B12 (µg)	3
EPA (g)	0.051
DHA (g)	0.26

How red you are!
Do I steal the show?

Most definitely.
Thanks. And thank you for inviting me.

You're welcome. But let me move on to more sensitive territory. Tell me about ciguatera.
How much do you know? And what do you wish to know?

Well, I know it's a neurotoxin present in fish that hang around coral reefs, such as, ahem, yourself. And that it can cause unpleasant symptoms like vomiting and diarrhoea in humans, and even blurred vision in some cases. But I don't know how widespread it is, nor whether you can spot it before it gets to you.

Yes, ciguatera comes from a certain type of algae, present in tropical and subtropical waters. Unfortunately, no, you can't spot it: it doesn't alter the aspect of the fish. And I'm not talking just about myself – it can also affect barracuda, grouper and others. But I need to point out that while the incidence of ciguatera is not negligible, symptoms tend to pass with antihistamines. There were five US deaths altogether in the whole of 2020, so we should put the whole thing into perspective.

Understood. To end on a more upbeat note: it's good to hear there are far more of you in the Gulf of Mexico than anyone thought.
Yes. I always knew it, of course. I just didn't want you humans to get too comfortable.

THE INTERVIEW
SNAPPER

Snapper tacos

TACOS DE HUACHINANGO

Mexico supplies almost two-thirds of the fish fillets sold in the United States of America. Domestic consumers are also well served: the country hosts the world's largest seafood market after Tsukiji in Japan. Complete with food stands and shops selling cookware, La Nueva Viga drains much of the Gulf of Mexico catch. But geared as it is towards the dense population basin around Mexico City, the market is unique by world standards in its deep inland location. Transportation costs mean Mexican fishing crews need to spend longer periods at sea to keep economically afloat. Some of the catch is directly exported north of the border.

Every day, but especially during big religious festivals such as Lent, La Nueva Viga fills with tens of thousands of shoppers. Piles of red snappers, filleted or whole, greet the eye from the stalls here. Both red snapper from the Gulf and its close Pacific cousin, *Lutjanus peru*, are known in Mexico as *huachinango*; the word is derived from the Náhuatl for "red flesh". Indeed, rather than the Gulf, the recipe featured here is from Baja California on the Pacific side, which supplies a tenth of the country's haul of *pescado y mariscos*.

Fish aside, this dish calls for either sour cream or mayonnaise (or, why not, both). If you go for the latter, it's much more rewarding to forgo the store-bought stuff and make your own. Away from this recipe, you can further customize your mayonnaise with chilli as an alternative to ketchup; with minced garlic for a sharp Mediterranean aioli; or else with chopped anchovies, capers and canned tuna – the dressing for the Italian 1980s classic *vitello tonnato*.

TO FEED **2 PEOPLE** YOU WILL NEED

2 SNAPPER FILLETS

1 LARGE TOMATO (OR 2 SMALL), MINCED

1 LARGE AVOCADO (OR 2 SMALL)

2 CUCUMBERS, SLICED

1 OR 2 GREEN CHILLI PEPPERS, DESEEDED, FINELY CHOPPED

1 HEAD CABBAGE, SHREDDED

2 LIMES (JUICED)

1 POT SOUR CREAM (AROUND 250 ML)

OLIVE OIL FOR FRYING

SALT AND PEPPER FOR SEASONING

FOR THE **TORTILLAS**

250 G FLOUR + MORE FOR DUSTING
•
2 TBSP VEGETABLE OIL
½ TSP SALT

FOR THE **MAYONNAISE**

1 EGG YOLK
•
1 TSP MUSTARD
•
MORE VEGETABLE OIL
•
A DROP OF LEMON

METHOD

1 In a bowl, toss together the tomato and chilli with lime juice, salt and pepper. Reserve.

2 Now make the tortillas. In another bowl, mix the flour, vegetable oil and salt, and pour 150 ml of hot water on top. Mix further, then knead on a flour-dusted board for a few minutes until you get a soft elastic dough ball.

3 Divide your dough into 6 parts and roll each one out with a pin, dusting with more flour. Heat up some vegetable oil in a pan and gently fry the tortillas one by one, on both sides, until golden. Remove, drain on kitchen paper, and cover to keep warm.

4 Make your mayonnaise (if using). Whip the mustard and egg yolk together and season with salt and pepper. Pour in the vegetable oil in a thin constant stream, whipping all the while to avoid splitting, until the mixture stabilizes into a semi-stiff, glowing yellow cream. You can use up to a third of olive oil as a substitute for vegetable oil, but too much of it will give your mayonnaise an unpleasant bitter edge. Finish with a drop of lemon, then cover and refrigerate to set further. (Lemon may also be used sparingly to "catch" the mayonnaise back if it splits.)

5 Score the skin on your fillets. Heat up the olive oil in a pan and fry the fillets on a medium flame, skin side down first. Remove from the pan and drain off the excess oil.

6 Taking care not to burn yourself, tear the snapper flesh and distribute it across the tortillas. Add the vegetables, sprinkle with the tomato and chilli mix, toss in a dollop of sour cream or mayo, and fold the stuffed tortillas over. Your *tacos de huachinango* are ready.

Fish mélange

The two-island country of Saint Kitts and Nevis forms an exclamation mark on the map of the Caribbean. The cuisine too – rooted in West African traditions and overlaid with British, French and South Asian influences – is vibrant, brash, straight-up. On the meat side, goat comes paired with breadfruit and green papaya.

Fish and shellfish are often grilled and laced with coconut, or, as here, served *en papillote* (i.e. in a parcel) with unfussy but colourful vegetables and mash. It is, in short, spunky comfort food to wash down with a brew rather than savour with fine wine – perhaps, if you can lay your hands on it, with a bottle or two of the local Carib lager.

I SMALLISH RED
SNAPPER, CLEANED

TO FEED
2 PEOPLE
YOU WILL NEED

I LARGE KNOB SALTED BUTTER

VEGETABLES:
OKRA, SWEET CORN
OFF THE COB, PEAS
AND DICED CARROT

1/2 ONION, SLICED FINE
+ 2 SCALLIONS (GREEN ONIONS)
CUT LENGTHWISE OR
IN 2-CM PIECES

SALT, PEPPER AND
A SPRIG OF THYME
FOR SEASONING

METHOD

1 Rub the fish with salt and pepper, inside and out. Insert the sprig of thyme in the cavity and set aside.

2 Rub a stretch of aluminium cooking foil with the knob of butter, and dot the remainder of the butter over the fish. Place the fish on the foil.

3 Cover the fish with the vegetables and the onions. Close up the foil and bake in the oven at around 175° C for 20 minutes or so, depending on the size of the fish.

4 Serve over mashed potatoes or yams, allowing the starch to soak up the juices. Kittian and Nevisian taste runs to spicy foods, so no one will blame you for adding hot pepper sauce on the side.

Trout

ONCORHYNCHUS MYKISS, SALVELINUS FONTINALIS, SALMO TRUTTA

Species of trout used to be geographically circumscribed: rainbow trout on the western edges of North America; brook trout in the Great Lakes region; brown trout in Europe. No more. Such has been the extent of global introduction for food and sport that these three main species – there are dozens altogether, subspecies included – can now be found in most latitudes. Breeding came early and energetically. As far back as the 1870s, in a novel instance of multipurposing public buildings, trout were reportedly being hatched from eggs in the basement of San Francisco City Hall. (The Great Earthquake and fire of 1906 killed 3 000 people and left City Hall in ruins. What happened to the fish, assuming any were left, history doesn't tell us.)

In the wild, trout live primarily in lakes and rivers. Depending on species and habitat, they can measure just 30 centimetres or up to four times that. Their weight might vary by a factor of 20. They'll run silver to gold in colour, or grey streaked with pink. You may find them covered in black or bright red dots. The alarmingly named cutthroat trout (*Oncorhynchus clarkii*) is stained crimson around its jaw and gills.

Remarkably, given this natural variety and global reshuffling, trout species remain distinct. Hybridization does occur in captivity, but far less in the wild, where it's been shown to curb reproductive fitness. Diversity within species can meanwhile be high. The brown trout swimming in British rivers alone are thought to be genetically far more diverse that the entire human race. Some brown trout may go to sea for a while, in which case they're called, intuitively enough, sea trout; most don't. And much like some human communities who'd sooner identify with brethren across the border, brown trout will snub other trout, but happily interbreed with salmon.

The salmon-trout kinship is well established: both belong to the Salmonidae family. Salmon is predominantly oceanic; trout, largely riverine. Consequently, trout's flavour echoes that of salmon as semi-skimmed milk echoes full-fat: substantially of a piece but subtler, as if holding back. At its best, trout tastes almost floral. Mild yet characterful, its flesh is a blueprint for versatility. Tomato, which features in two of our recipes, is a frequent partner: its acidity sets off the sweetness of the fish. Yet trout will easily accommodate more adventurous companions. In the French city of Reims, in Champagne country, Chef Kazuyuki Tanaka has been known to pair the Arctic char (*Salvelinus alpinus*) – a trout found in the cold lakes of Europe and Canada – with fermented fennel or cocoa. Indeed, something about trout seems to incite flights of fancy. One contemporary source has the nineteenth-century composer Franz Schubert improvising his most famous Lied, *Die Forelle* (The Trout), after partaking liberally of Hungarian red wine. In Schubert's song, the frolicky trout, exhilaratingly free in the first two stanzas, is subdued by a ruthless angler in the third. Musical audiences may squirm at this turn of events. Gourmands, though, will rejoice: for them, trout's most beguiling note is the one struck post-mortem.

One fish, many guises

KNOW
YOUR FISH

Trout is massively sourced from aquaculture these days. Rainbow alone comes in at around 850 000 tonnes a year, a near-twentyfold increase since the early post-war era. The Islamic Republic of Iran and Türkiye take the top spots; Chile and Peru, which farm some of their trout in seawater, are also big producers. Norway, another sizable supplier, labels its variety "fjord trout". Food sector aside, trout remains a popular game fish and, in many countries, a boost to rural economies. Every year, significant quantities of farmed animals are used to restock lakes and rivers whose wild populations have dwindled.

This relative abundance makes trout a fairly inexpensive food – so much so that in France or Italy you may find it sold as *truite saumonnée* or *trota salmonata*, in a belief that drumming up an association with salmon will enhance its prestige. In fact, this just means that the flesh has been made redder through the inclusion in the feed of synthetic astaxanthin. (The addition of astaxanthin, a substance found naturally in krill, is a safe and legal practice, although its effect is essentially cosmetic.)

Alongside a high omega-3 count, trout delivers a good protein kick for relatively few calories. It's also rich in selenium, an essential antioxidant. Steamed, grilled or oven-roast, this is a fish that's at its best in pared-down company: overcomplex garnishes might smother it. Consequently, the recipes in this chapter keep things simple (though the Kyrgyz dumplings may take a bit of mastering). As usual – but even more so in the case of trout – make sure you don't overcook your fish, or it will desiccate unpleasantly. One US rule of thumb suggests baking trout for 10 minutes per inch (2.5 centimetres) of thickness in the case of fillets, or a couple of minutes more if the fish is on the bone or wrapped in foil.

Nutrition facts

RAINBOW TROUT, FARMED, FILLET, RAW (NORWAY)
per 100 grams

ENERGY (kcal)	196
PROTEIN (g)	19.3
CALCIUM (Ca) (Mg)	9
IRON (Fe) (Mg)	0.3
ZINC (Zn) (Mg)	0.4
IODINE (I) (μg)	12
SELENIUM (Se) (μg)	20
VITAMIN A (RETINOL) (μg)	18
VITAMIN D3 (μg)	7
VITAMIN B12 (μg)	3.9
OMEGA-3 PUFAS (g)	3.18
EPA (g)	0.81
DHA (g)	1.37

What are you humming?

Hold on, I'm meant to ask
the first question...
WHAT. ARE. YOU. HUMMING?

It's a famous Lied named
after you. A Romantic piece.
It was popular in Europe
in the early nineteenth
century.
Really? Why wasn't I informed?

I wouldn't know. Maybe
because it doesn't end well
for you.
Did it end well for the composer?

I'm afraid not. He died
young. He was only
31 years old.
Well, poetic justice.

Don't be macabre. People
associate you with joy and
vivaciousness.
But they still respect me less than
salmon?

I'm not sure what to say to
that. It's true, you're seen
as more common.
I suppose you're a victim
of your own success:
you've been introduced all
over the world.
Well, I guess I'm highly adaptable.
It's one of my main qualities. Also,
I'm mostly a freshwater fish, but I
don't taste muddy, as they tend to.

You know, this thing about
freshwater fish tasting
muddy: it's a bit of a myth.
There may be a peculiar
flavour, but a lot of the time
it's about lack of freshness.
There is no connection to
mud in any case.
Whatever.

Let me just say that you're
looking rather resplendent.
Not technically like a
rainbow, but I can see why
they call you that.
Thank you. Make sure salmon
knows!

THE INTERVIEW
TROUT

Baked Sevan trout

ISHKHAN, իշխան

Sitting at the junction of the Islamic Republic of Iran and Türkiye, the world's two leading trout producers, Armenia is home to its own highly prized species, ishkhan (*Salmo ischchan*). The fish, endemic to the waters of Lake Sevan, a historically much-tormented ecosystem, is endangered; two subspecies are already extinct. In the late 1970s, commercial exploitation of wild-caught ishkhan was banned by the then-Soviet authorities, and the lake area declared a national park. If you're having ishkhan in Armenia, it should be certifiably farmed. Elsewhere, any other kind of trout will make a perfect substitute.

The use of pomegranate here is an elegant, unmistakably Caucasian touch. It's worth securing the fresh fruit, fairly widespread these days, or else buying the bottled seeds.

**I WHOLE TROUT,
I KG OR SO, CLEANED**

**TO FEED
4 PEOPLE
YOU WILL NEED**

METHOD

2 TBSP TOMATO PASTE

2 TBSP SOUR CREAM

**HALF A POMEGRANATE OR
THE BOTTLED EQUIVALENT
OF FRESH SEEDS**

I TBSP PAPRIKA

SALT

1 Preheat the oven to 200° C. Meanwhile, make diagonal incisions into the fish 3 cm apart.

2 Combine all the ingredients except the pomegranates. Rub this marinade all over the fish, inside it and into the incisions.

3 Roast the fish in the oven for up to half an hour or so, depending on size. While the fish is cooking, toss out the pomegranate seeds by beating the outside of the fruit with a spatula and giving it a little squeeze. (Ensure any white pith, which is bitter, is discarded.)

4 Serve the fish sprinkled with pomegranate seeds, accompanied by long grain rice. For a proper Armenian feast, add toasted pine nuts and almonds to the rice.

Trout in tomato sauce

TRUCHA EN SALSA DE TOMATE

Wedged between the Atlantic and the Pacific, Costa Rica, with its coastal delights, is rarely associated with freshwater pursuits. But the cold streams of the interior host populations of trout that attract a large number of fly-fishers. Aquaculture is also well developed, and while rainbow trout comes a distant third in the rankings – tilapia heavily dominates the sector – production has doubled in a decade. The trout in our recipe is set off by a profusion of herbs, making this a zesty counterpoint to the heartiness of our Kyrgyz offering on p. 141.

TO FEED
2 PEOPLE
YOU WILL NEED

2 TROUT FILLETS

1 CLOVE GARLIC, CRUSHED

3 TOMATOES

5 SPRIGS OREGANO, FRESH OR DRIED

SALT, 1 TSP CRUSHED BLACK PEPPER AND ½ TSP CRUSHED CUMIN SEEDS

OLIVE OIL FOR FRYING

3 LEAVES FRESH BASIL

METHOD

1 In a pan, warm a couple of spoonfuls of olive oil and turn down the heat. Season the fish fillets with salt, pepper and some oregano and gently sauté the fillets until coloured and tender.

2 Meanwhile, warm more olive oil in a separate pan. Chop the tomatoes and throw them in, adding the torn basil leaves, the cumin and the remaining oregano. Add a pinch of salt, and also a little sugar if the tomatoes are too sharp. Stir and reduce over a low flame to a thick sauce.

3 Serve the warm *trucha* with the tomato sauce, spooning over any cooking juices, alongside brown rice, green beans and cold shredded red cabbage.

Fish manty

BALYKTAN ZHASALGAN MANTY, БАЛЫКТАН ЖАСАЛГАН МАНТЫ

Kyrgyzstan is described as the country furthest away from any ocean. By way of natural compensation, its river system is stupendous: tens of thousands of waterways, mostly fed by glaciers, run through Kyrgyz mountains. The lakes number close to 2 000. Issyk-Kul, by far the largest of these, saw the introduction of ishkhan trout (see p. 136) from Lake Sevan in Armenia in 1930: so well did the fish take to its new environment that it reportedly grew, in some cases, to five times its original size. Trout, in other words, is at home in Kyrgyzstan, and the country is an emerging destination for sports fishing. Fish production, however, fell precipitously after the collapse of the Soviet Union. Since the trough point of 2009, FAO's technical assistance has helped the sector bounce back: by 2021, there'd been a fourteenfold jump in output.

In our Kyrgyz recipe, the trout is finely chopped and married to potatoes and onions. This, satisfying as it is, may not be the most unexpected of pairings. But the charm of the dish lies in the *manty*, the famed Turkic dumplings which encase the mixture. In Kyrgyzstan, as elsewhere in central Asia, a smear of lamb fat (known in Kyrgyz as *kurdyuk*) might be added to each *manty*: you won't taste it, but it will make for a luxurious mouthfeel. Do that, by all means, or else use butter – or indeed, nothing at all if these things aren't part of your diet. On the other hand, you may wish to chop parsley or dill into the stuffing for a fresher kick.

In the region, the dumplings are steamed in a special dish called a *mantyshnitsa*, but a bamboo steamer of the kind used for dim sum will do just as well.

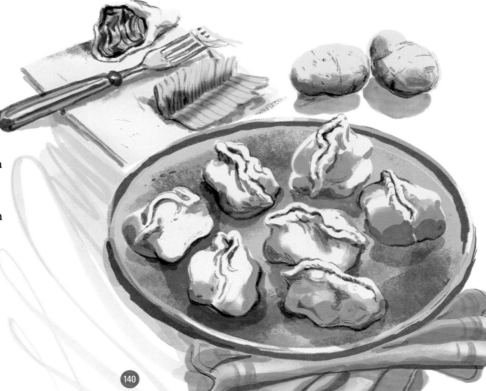

STARTER FOR
6 PEOPLE
YOU WILL NEED

TO FEED
4 PEOPLE
YOU WILL NEED

1 RAINBOW TROUT, UP TO
1 KG, CLEANED, SKINNED AND
THOROUGHLY DEBONED

1 LARGE POTATO
OR 2 SMALLER ONES

PARSLEY OR DILL
(OPTIONAL)

2 ONIONS

SALT, RED ALEPPO PEPPER
AND BLACK PEPPER TO TASTE

FOR THE
MANTY

500 G WHITE FLOUR

1 EGG

SALT TO
TASTE

METHOD

1 Start with the *manty* preparation. Sift the flour onto a chopping board, then shape it into a mound and make a crater in it. Break the egg into the flour crater and add a pinch of salt. Using a fork, incorporate the egg into the flour, taking care to avoid the formation of lumps. Adding a little water at a time, knead the mix into a firm dough and let it rest.

2 Now make the mince. Peel the potatoes and onions and dice them up, together with the skinned, boneless fish. Season with salt and pepper, add any herbs if using, and whizz the lot lightly in a blender. (You're looking for a paste that still has some structure, not a homogenous mush.)

3 Turn your attention back to the *manty*. Sprinkle some more flour onto your chopping board. Taking chunks out of the dough, roll it out as thin as you can with a rolling pin, trying to ensure an even surface. Using a sharp knife or round pizza cutter, cut out squares of about 8 cm per side.

4 Smear some fat onto each dough square if using, then place a dollop of fish mince in the middle. Now gather up the corners of each square to form a peak and twist the peak slightly to form your manty. Press the seams together to make sure the parcels don't break during cooking.

5 Steam the *manty* for 15–20 minutes over boiling water. Taste one to check they're done, adjust the cooking time as needed and serve.

Tuna

THUNNUS

....

Tuna is mackerel's cousin – the cousin who grew to be bigger, race faster and smell more refined. From the subclan's humblest member, the half-metre-long bullet that is mackerel's nearest cognate, to the bluefin going for top dollar at Tokyo's boisterous fish auctions, tuna, like mackerel, are Scombridae – as is the bonito, included here for ease of reference. Indeed, at the edges of the tuna constellation, identities become somewhat blurred. In some taxonomies, and particularly for commercial purposes, the bullet tuna is in fact a mackerel; while the bonito is, in effect, a tuna.

Present in various incarnations across fishing zones, tuna are apex predators. They feed on many fellow marine creatures, mackerel included; by contrast, few others feed on tuna. These hunters live long, swim fast and migrate far. Their muscular flesh runs to deep pinks and crimsons; the skin is like sun-baked aluminium, streaked with greyish blues.

Tuna are large – often *very* large – fish, which means they're never eaten whole. Much as for the big land animals that form part of modern diets, cuts run the gamut of desirability. The marbled, exquisitely flavoured, fattier parts from the belly of the fish – known as *ventresca* in Italy and *otoro* in Japan – tend to be the most prized: they are to tuna what Kobe fillet is to beef.

Cuts aside, different species have different uses. Prices rise in line with prestige (though it's fair to say these hierarchies have varied with the decades: as with lobster, for example, shifting fashions have moved erstwhile workaday species up the value chain). At entry level, we find the stuff that supplies most markets with canned tuna. Best friend to students, singles and Sunday cooks, this is generally the pale albacore in its superior version, and a slightly rougher, oilier skipjack at the more democratic end. Thailand is the world's top canned tuna exporter, cranking out billions of dollars' worth in 2020.

Yellowfin, the next step up on the marketing ladder, is a versatile fish, available as steak or sushi; bigeye tuna provides a higher grade of the

142

latter. Both are known in Hawaii and the wider United States of America as *ahi* – and dominate the canned segment in Italy. Most exclusive of all is bluefin, whose lush, gamey flesh winds up as the choicest sashimi: here, the raw fish is given star billing, without the admixture of rice or vinegar. Japan alone concentrates four-fifths of the market.

With tuna being the highly migratory fish that they are, roaming across multiple jurisdictions, regional fisheries bodies have been set up to issue binding rules for sustainable management. Because of its premium appeal, bluefin has long concentrated efforts. Pacific Small Island Developing States (PSIDS), for example, have teamed up to cap fishing licenses. Meanwhile, "ranching" of bluefin – that is, catching young individuals and raising them in captivity – has been spreading since the early 1990s. Farming bluefin from eggs, long thought a technical impossibility, is also shaping up as a viable option.

Such practices have somewhat relieved pressure on wild stocks. In 2021, the International Union for Conservation of Nature (IUCN) declared Pacific bluefin "of least concern" (though still severely depleted, at less than 5 percent of its original biomass) and Atlantic bluefin "near-threatened" – an improvement for both. Southern bluefin is endangered, though no longer critically so. Overall, ethically minded buyers may wish to approach bluefin with a discerning mind.

Racers of the seas

KNOW
YOUR FISH

Wholesale buyers – or their agents – who purchase tuna off the boat will grade the catch along a scale running from 1 (highest) to 3 (lowest). Some systems have as many as eight steps, from minus 1 to 3 plus, with much nuance around the 2 mark. The grading is generally done by cutting crescent-shaped slivers from the tail end of the fish and extracting spindly "plugs" from its core. The slivers and plugs are then laid out in pairs (one sliver, one plug) on white Styrofoam and examined in natural light. Characteristics such as the degree of translucency, and the intensity and consistency of colour, are assessed. Graders will also look for any indication of "burn" from lactic acid or for evidence of disease, which can cause discolouring or make the flesh spongy to the touch.

Much of this millimetric dance, not unlike the judging of a wine's robe, will elude the untrained eye. As a retail buyer, canned tuna aside, you're most likely to encounter slabs or chunks of yellowtail of varying quality. And even without a grader's eye, there's much you can tell.

To start with, beware improbably low prices. Tuna is a costly fish – some estimates suggest it pulls in USD 40 billion globally, once all additions along the value chain are factored in – and thus prone to fraud and shady dealings. Also, mind the colour. In developed markets such as the European Union, artificial colouring is outlawed. Check for signs of lapsed freshness in the form of opaque spots or dull brown patches.

In the European Union, as in all other cases for fish caught in the wild, the FAO fishing zone must be mentioned. Retailers should also make it clear if the tuna is fresh or previously frozen. In some countries, the grade will often be specified. If so, anything 2+ or finer should be a fair bet; the closer to 1, the more you can afford to have the fish very rare. Conversely, tuna offcuts may be had cheaply, especially at the end of market day. These are frequently streaked with hard white sinew, which is chewy and best avoided in anything approaching raw form. Save the offcuts for a stew.

If your fish is labelled *ahi*, you may want to dig further. Bigeye is classified as vulnerable; yellowfin is not. If buying raw fish, you can sometimes tell yellowfin by the presence of a dark T-shaped bloodline.

Finally, picking through your bluefin tuna is something of an ethical and epistemological tightrope. Pacific bluefin is now, broadly speaking, fair game. It's also a vital source of income for a small nation such as the Marshall Islands, where a tenth of the workforce makes a living by fishing it. Atlantic bluefin is more fraught. While the threat to overall stocks has been downgraded, this is largely down to some recovery among Mediterranean populations; in the western Atlantic, the species remains endangered. To add to the complexity, mislabelling of bluefin is rife, and many restaurants still shun it.

Nutrition facts

TUNA, FRESH, YELLOWFIN, RAW
per 100 grams

ENERGY (kcal)	109
PROTEIN (g)	24.4
CALCIUM (Ca) (Mg)	4
IRON (Fe) (Mg)	0.8
ZINC (Zn) (Mg)	0.4
SELENIUM (Se) (μg)	91
VITAMIN A (RETINOL) (μg)	18
VITAMIN D3 (μg)	2
VITAMIN B12 (μg)	2.1
EPA (g)	0.012
DHA (g)	0.088

Bluefin tuna! Finally, we meet. It's an honour.
Who? What? Who are you?

I sent you an interview request. You accepted.
Did I? Okay. I can't stay long.

Oh? I'm sorry that you're in such a hurry.
Yes. Always. I have to keep moving. They've said about me that I'm all heart. It's almost true: my heart is vast. Needs constant pumping. I never rest. If I stop, no oxygen for me.

I see. I think some sharks are like that too. And I'm told there's also something called countercurrent exchange. Would you care to explain?
It's an instance of *rete mirabile*, or "wonderful net". It's like this: I have two bloodstreams. My blood flows in opposing directions. But these bloodstreams intersect. They exchange heat. This elevates my body temperature.

And this thermoregulation allows you to swim and hunt more efficiently, right?
Correct. I'm up to 15 °C warmer than the water around me.

Remarkable. I'd now like to broach the complex issue of your relationship with mackerel. In some ways, mackerel is part of your family...
Pfff... Impostor. There's only one way to deal with it.

Which is?
Eat it.

Ahem... All right. Yes, I suppose we all do. But we eat you too.
No comment.

Okay, I respect that. Looking back, I may not have had answers to all my questions, but writing this book has been immensely instructive.
Good. Send me an autographed copy. I must dash now.

THE INTERVIEW
TUNA

Fish stew

QELYEH MAAHI, ماهی قلیه

With its traditional cuisine leaning heavily towards meat (lamb in particular), the Islamic Republic of Iran holds a middling rank in the fish consumption leagues. Even so, per capita intake has been growing steadily, and the country is turning into a fisheries powerhouse. The blessing of an extensive double shoreline – along the Persian Gulf and the Gulf of Oman in the south, and the Caspian Sea in the north – has a lot to do with this bonanza. The industry employs more than 200 000 people; fish production doubled between the mid-2000s and the mid-2010s. But much of the credit also goes to a booming aquaculture sector, whose output had topped a million tonnes at the time of writing.

This recipe draws on the gastronomic culture of Bushehr: the southern province, a coastal strip facing Kuwait across the Gulf, is known for its seafood cuisine. *Qelyeh Maahi* calls for tuna or mackerel, depending on the versions and the degree of refinement sought. Other fish may do, though none too bland or flaky. You'll want a dense texture to withstand the long simmer, plus enough intrinsic flavour to carry off the spicing.

Tamarind is rich in vitamins B1 and B3 and in potassium. It should be easy enough to find these days, either whole in its hard shell, or else peeled and pressed into bricks. The pulp needs loosening in water; syrupy and tangy, it will lend your stew a delightfully tart note. All told, the finished dish should make you feel you've alighted at the gastronomic midpoint between India and the Balkans. Which is, of course, exactly where Bushehr lies on the map.

If no tamarind can be found, grab some plump dates instead and mash them with lime juice. Strain this mixture as you would the tamarind.

50 G TAMARIND

2 THICK TUNA STEAKS

(around 350–400 g.
You may want to cut these
further into halves.)

TO FEED 2 PEOPLE YOU WILL NEED

**I SMALL ONION, OR
HALF A LARGE ONE,
CHOPPED**

**2–3 CLOVES
GARLIC, FINELY
CHOPPED**

**I BUNCH CORIANDER
(CILANTRO), CHOPPED**

**OPTIONAL:
A PINCH OR TWO OF
SAFFRON, ADDED
RIGHT AT THE END**

**I TBSP TO 25 G FRESH
FENUGREEK LEAVES,
CHOPPED – OR SEEDS
IF UNAVAILABLE**

**A COMBINED FISTFUL
OF SALT, BLACK PEPPER,
TURMERIC, CORIANDER
SEEDS AND CHILLI FLAKES**

METHOD

1 Soak the tamarind pulp in 100 ml of warm water. Leave to stand for 30 minutes to an hour, then mash well with fingers or a fork. Strain the mixture through a sieve. You're aiming for a fluid paste – if needed, thicken with a little flour (or possibly honey if you like your food less sour).

2 While the tamarind is infusing, grind the salt, pepper and spices together. Use half this spice mix to rub the fish on all sides, adding a little water. Leave the fish to absorb the spiced marinade. Reserve the other half of the spice mix.

3 Separately, chop the onion and sweat it gently in oil until soft. Add the reserved half of the spice mix and stir well. Simmer until dark golden in colour.

4 Add the garlic to the pot. Fry the lot for another minute, taking care not to burn the garlic, then tip in the chopped coriander and fenugreek (or fenugreek seeds). Cook for a further 10–15 minutes, adding a little water to keep the sauce loose, until the content of your pot turns emerald green.

5 Pour in the tamarind paste, then another glass of water. Continue to reduce the sauce. Taste and add more salt if needed.

6 Separately, flash-fry the tuna pieces in a hot pan. While the inside is still pink, remove them from the pan and add them to the pot, basting them with the green sauce.

7 Let the *qelyeh maahi* bubble away on a low flame for 20 minutes to half an hour, partly covering the pot and giving it the occasional shake. Don't stir the pot: the fish – or chunks of

it – should remain whole. Add more water in small quantities if necessary. Keep tasting to make sure the coriander's bitter edge has been cooked out.

8 If using saffron, add a couple of threads shortly before serving. Give them a little local jostle to spread the flavour, then turn off the flame and leave to stand for a minute. Plate up on top of rice, with pickled vegetables on the side.

Lemon fish soup

MA' SOURA, معصورة

Oman's spectacular coastline cleaves the Gulf. Despite increasing wealth, the country retains some of the slower, artisanal modes of its rustic past. The national fishing fleet consists overwhelmingly of traditional boats; industrial vessels supply just a fraction of the catch.

Fish production has been rising year on year, with a fair share of exports going to Asian markets. This dynamism ties in with broader GDP growth. But there's another, less upbeat and more incidental factor: Oman has reported to the Indian Ocean Tuna Commission that bigger hauls also correlate with a "slowdown in fishing pressure" across the border, in conflict-ridden Yemen.

Oman's location astride trade routes accounts for the Indian, Persian, African and eastern Mediterranean influences on its cuisine. Yet a frugal Bedouin heritage also means that tasty simplicity, rather than showy opulence, is the norm. In our example, the fish is grilled, then shredded into a hearty, acidic, spicy soup. The recipe uses a whole Red Sea tuna: it's meant to feed a large family and, true to this convivial informality, lists no prescriptive quantities. As most readers, fisherfolk aside, are unlikely to have a whole tuna at their disposal, now's the time to use those cheaper tuna cuts, or – again – go for mackerel instead.

Lemon gives this dish the sourness that is a hallmark of regional cuisine. Dried za'atar leaves can be found in Near Eastern food stores or speciality spice shops: substitute oregano, thyme, marjoram or (even better) a mixture of all three if za'atar proves elusive.

TO FEED
A LARGE
FAMILY
YOU WILL NEED

RICE

I SMALL TUNA, OR TUNA CUTS*

TOMATO, CHOPPED

PARSLEY, FINELY CHOPPED

GREEN ONION
(SCALLION) – THE
GREEN PART ONLY,
FINELY CHOPPED

ZA'ATAR

CUMIN POWDER

SQUEEZED LEMON OR
LIME JUICE

GARLIC, CHOPPED
OR GRATED
•
GREEN CHILI
AND CHILI FLAKES
•
SALT

*The recipe recommends cooking the fish on the bone, which augments the flavour, so in the absence of a whole tuna, consider using mackerel.

METHOD

1 Roast the fish (previously cleaned and gutted) thoroughly in the oven. Alternatively, grill it.

2 When the fish is done, leave it to cool, then remove the flesh from the bone, preferably by cutting it to preserve whole chunks for the most part. You will end up with flakier bits too, but these will add texture to the soup.

3 Separately, boil the onion leaves till soft. Drain and reserve.

4 Place all the ingredients and fish in a saucepan, cover with water, and leave to simmer, allowing the water to reduce. Salt as needed. When the soup is still a little more liquid than you'd like it (the rice will thicken it further), turn off the heat and cover to keep warm.

5 Boil the rice, then drain it and sauté it quickly in a frying pan with the cumin and chopped garlic. When fragrant, combine with the soup and serve. You can garnish the *Ma'soura* further with ginger, coriander and lime.

Cebiche Jipijapa

Despite a comparatively modest length of coastline, Ecuador is the top tuna-fishing power in the eastern Pacific: it pulls in over a third of the regional haul. Much of the processing and canning function is concentrated in the province of Manabí, with the port city of Manta seen as the nation's tuna capital.

While it has yet to achieve the heights of recognition enjoyed by Peruvian cuisine, Ecuador's riffs off a similar mix of Indigenous, western and modern Japanese traditions. Contemporary takes on the national fare flourish in Quito and Guayaquil as they do in Lima. Root vegetables and varieties of corn unknown elsewhere fuel this sense of adventurous rediscovery. In Ecuador as in Peru, and especially in coastal areas, cebiche (the *b* spelling is preferred locally) cuts across the folk-urban divide. The technique of "cooking" raw fish in a cold citrus marinade is the same; it's the stuff around the fish that tends to vary. Ecuadorean iterations are generally soupier – with more of the marinade retained or spooned over – and less spicy. In Ecuador, tomatoes (and even ketchup) are commonly added, where more purist versions tend to dominate in Peru. And while in Peru, the dish is frequently served with corn or spuds, Ecuadoreans will often pair it with green plantains.

This version, peculiar to the town of Jipijapa in Manabí province, uses bonito – a more affordable fish that hovers biologically at the edges of the tuna family: it is virtually indistinguishable from skipjack. Bonito is popular around the world, from the Black Sea to the Pacific. In Japan, where it's known as *katsuo*, bonito is fermented, dried and flaked to make the condiment *katsuobushi*: looking rather like pencil shavings and scattered over warm food, the flakes dance prettily in the air before settling into a smoky, rose-coloured mess.

In Ecuadorean cebiche Jipijapa, the bonito is accompanied by *mantequilla de maní* (peanut butter). As per this book's rule of thumb, you can use the store-bought stuff or, more heroically, make your own. The recipe is included.

Patacones are plantain chips. For some of you reading this, securing the plantains may be the toughest job here. If you can't get them, don't be tempted to replace them with sugary bananas: opt for some quality ready-made chips instead, such as blue corn or root vegetables. But if you do get plantains, slice them up as you would a banana, salt them a little, dust them with flour and fry them in vegetable oil till golden and still fairly soft. Drain off the excess oil, then place the slices between two sheets of kitchen paper, and use a spatula or the bottom of a jar to flatten them into thin discs. Refry the slices of plantain till crisp, drain them again, and you're done.

TO FEED **4 PEOPLE** YOU WILL NEED

1 KG BONITO, DICED, LEAN AND CLEAN PIECES ONLY

1 DOZEN LIMES

(Ecuadoreans would use a citrus variety called limón sutil, but limes will do perfectly.)

1 LARGE TOMATO, OR 2 SMALL ONES, AND SOME TOMATO CONCENTRATE

1 ONION, CHOPPED

FRESH CORIANDER (CILANTRO)

1 AVOCADO, DICED

OLIVE OIL

MUSTARD

COARSE SEA SALT

FOR THE **MANTEQUILLA** DE MANÍ

1 HANDFUL RAW PEANUTS
•
½ TSP SUGAR
•
⅓ TSP SALT

METHOD

1   Cover the fish with the lime juice and let it marinate for 2 hours.

2   Separately, make the *mantequilla de maní*. Dry-roast the peanuts in a frying pan, stirring constantly to avoid burning, until dark and fragrant. Transfer to a food processor and whizz with the olive oil, sugar and salt until you obtain a rich, semi-solid paste.

3   When the fish has coloured and lost translucency, divide it into 4 portions and place each portion onto a deep plate. (Glass plates or large glass bows will enhance the appearance of the dish.) Spoon the lime marinade and some olive oil over the fish.

4   Around the edge of the plate, add the remaining ingredients: the diced avocado, chopped onion and tomato, a squeeze of tomato concentrate, a dollop of mustard and one of peanuts. Garnish with the fresh coriander and the coarse sea salt.

RIVER FISH
OF THE AMAZON BASIN

Over stretches of time that most of us readers and writers of cookery books would scarcely comprehend, the Amazon has shuffled and reshuffled its surrounding landscape, forming lakes where water once flowed and causing myriad river creatures to branch off genetically. Then again, riverbanks crumble as often as they rise. Lakes are reversed no less than created. Having split off, the fauna re-mingles, an evolutionary to-and-fro that has endowed the Amazon basin with astonishing biodiversity. In the lands that are now Brazil and neighbouring countries, thousands of endemic fish species have sustained Indigenous Peoples for millennia. In recent centuries and decades, the creolization of human populations and the expansion of gastronomic frontiers have de-enclaved some of these fish, bringing them to tables further afield. Yet overall, the fish of the Amazon are of that place and of nowhere else. When, early in 2021, a pirarucu was found floating in the waters off Florida, it was feared the southern US state had to contend with yet another invasive species. But it's far more likely the fish had been kept as an exotic pet and was discarded there lifeless.

Pirarucu

ARAPAIMA GIGAS

Back in the late 1990s, pirarucu (*Arapaima gigas*), one of the world's largest freshwater fish, was thought close to extinction. Twenty years later, thanks to strict management programmes involving Brazil's Indigenous communities, stocks had soared nearly tenfold to almost 200 000 individuals. Fishing of the pirarucu – a Tupi word; the fish is known across the border in Peru as *paiche* – is only permitted in the second half of the year, the non-mating season. The centralization of sales through a cooperative also means local fisherfolk are paid twice per kilogram what they would raise from selling directly to local markets. Even so, poaching remains rife, and connected to broader criminality and environmental destruction.

A carnivorous predator, pirarucu is an oddity on multiple counts – in size certainly, coming in at up to 3 metres and 200 kilograms (though more commonly found at around half that). But also in a number of quasi-fantastical traits: it has a python-like body; a tongue studded with teeth; and a capacity to breathe by swallowing air, like humans. This exceptionalism echoes through Indigenous legends that cast the pirarucu as a warrior, given piscine form in defeat.

KNOW YOUR FISH

Seen whole, the pirarucu looks semi-reptilian. The head is bony and sits small on the body: it appears anatomically mismatched, in the way children's drawings are.

Underneath a sheath of piranha-proof skin, peppered with red, the flesh has a pleasing rose tinge.

Analysis of the pirarucu's dorsal and ventral muscles by Brazilian nutritionists has revealed the presence of 27 beneficial fatty acids – and with stocks back at healthy levels, the fish has cropped up on hip restaurant menus in Brazil's coastal cities. This remains, that said, a highly local species: it is essential that it be cooked in its homeland, with guaranteed traceability; or, if replicating pirarucu dishes abroad, that it be replaced with a sustainable alternative. Cod (see p. 50) would do well instead: although a river animal, the flavour of the pirarucu is more akin to that of oceanic equivalents. Early European settlers, in fact, dubbed the pirarucu "Amazonian cod". They found it lent itself to salting and drying just as well as *bacalhãu*.

Nutrition facts*

PIRARUCU, CULTIVATED IN RONDÔNIA, BRAZIL. SEVERAL CUTS, FRESH (average of fish with different weights and cuts)

per 100 grams

ENERGY (kcal)	106
PROTEIN (g)	20.6
CALCIUM (Ca) (Mg)	13.1
IRON (Fe) (Mg)	0.1
ZINC (Zn) (Mg)	0.6
VITAMIN A (RETINOL) (μg)	TR
VITAMIN B12 (μg)	1.4

Nutrition facts for other Amazonian river fish species N/A

Amazonian fish stew

MOQUECA AMAZÔNICA

Moqueca harks back to the *mu'keka*, an Angolan dish that may be compared to a ragoût or fricassée. This lineage – *moqueca* is closely associated with Salvador de Bahia, the Brazilian metropolis whose residents are mostly of African descent – likely accounts for some similarity with the seafood gumbo of Louisiana. Distinctly Afro-Brazilian is the use of *dendê*, the red palm oil that gives a soulful warmth to much Bahian cuisine. If you have access to sustainably sourced Brazilian *dendê*, do use it; if not, olive oil will do. For the fish, we suggest adding one part haddock, possibly smoked, to ten parts pirarucu/cod. This turns the haddock, in effect, into a solid fish condiment – for fish.

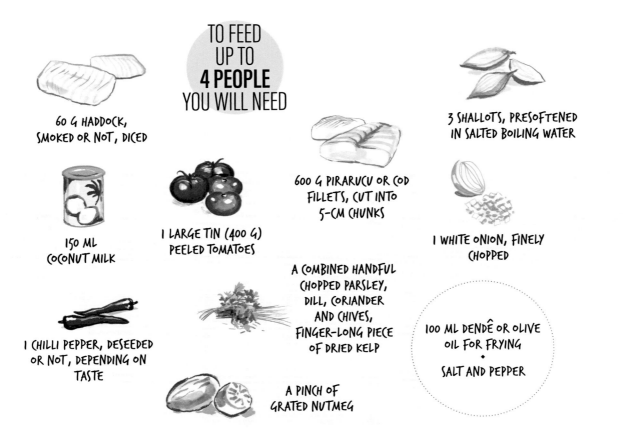

TO FEED
UP TO
4 PEOPLE
YOU WILL NEED

60 G HADDOCK,
SMOKED OR NOT, DICED

150 ML
COCONUT MILK

1 LARGE TIN (400 G)
PEELED TOMATOES

600 G PIRARUCU OR COD
FILLETS, CUT INTO
5-CM CHUNKS

3 SHALLOTS, PRESOFTENED
IN SALTED BOILING WATER

1 WHITE ONION, FINELY
CHOPPED

1 CHILLI PEPPER, DESEEDED
OR NOT, DEPENDING ON
TASTE

A COMBINED HANDFUL
CHOPPED PARSLEY,
DILL, CORIANDER
AND CHIVES,
FINGER-LONG PIECE
OF DRIED KELP

100 ML DENDÊ OR OLIVE
OIL FOR FRYING
•
SALT AND PEPPER

A PINCH OF
GRATED NUTMEG

METHOD

1 Season the chunks of pirarucu or cod with salt, pepper and nutmeg. Roll them in grated garlic, half the chopped onion and half the chopped herbs, and leave them in the fridge for an hour.

2 In a large pan, sauté the diced haddock with the remaining chopped onion until the onion becomes translucent. (This is your fish-enriched *refogado*, the Portuguese term for a base of allium fried in fat.) Add the seasoned pirarucu and continue to cook on a medium flame for 3 minutes or so, until golden.

3 Tip in the peeled tomatoes and coconut milk, adding the chilli pepper. Fork through the tomatoes to help them disintegrate. Bring the mixture back to a simmer, then add the softened shallots. Taste and adjust for salt if needed.

4 Simmer the moqueca for another 3 minutes or so, then remove from the flame. Sprinkle the remaining herbs on top and serve with rice or other starchy accompaniment – polenta or a Brazilian manioc-based *farofa* (for a wider discussion of manioc, see the recipe for Guyanese tambaqui tuma pot on p. 161).

157

Tambaqui

COLOSSOMA MACROPOMUM

Tambaqui is arguably the premier Amazonian food fish. Looks-wise, it is large and flat, up to a metre in length, and shaped like a big fleshy diamond. Now introduced beyond the Amazon, it can be sourced from aquaculture abroad – though not widely at the time of writing.

Barely qualifying as an omnivore, the tambaqui mostly feeds on fruit and nuts: it gulps them down from the water surface, then repays the kindness by dispersing their seeds as it excretes. The odd snail and insect rounds off its diet.

In the wild, the tambaqui prefers the clear waters of flooded forests – until it's time to spawn, when it likes things rougher and migrates to whitewater rivers. This is, however, a fairly adaptable fish, tolerant of environments that are poor in oxygen and even mildly saline. Its ecology may depend on the habitat, and its colour too.

KNOW YOUR FISH

Tambaqui can be grey, or yellowish, or even reddish as glimpsed at fishmongers' stalls (most likely in Manaus, the Brazilian city that is the main market for Amazonian foods). They may also sport a dark splotch that wraps around the abdomen and runs up the sides, a feature typical of a subspecies known as black pacu. The eyes are bulgy, the teeth sharp. The flesh is white and firm, ideal for grilling, with a sweetness widely credited to the fish's fruit-based diet. Sea bass (see p. 118) would make a good substitute.

Tambaqui tuma pot

Our tambaqui recipe taps into the Indigenous tradition: tuma refers to the cooking of meat or fish in *kadakura*, the Guyanese name for cassava water. North American readers may know cassava as yuca or Brazilian arrowroot; French speakers will recognize it as manioc. It is, in many developing countries, a reliable source of carbohydrates. It's also easy enough to obtain these days, even in regions where it isn't native. But the need to soak the cassava to eliminate toxicity (it contains traces of cyanide), then grate it and strain it to extract the water – Guyana's Indigenous Peoples would use a flexible cylindrical basket called a *matapi* for this – means you're better off attempting the tuma when you have a) time and energy to spare, and b) enough mouths to feed to make it worth it. True to this logic, our recipe lists quantities for eight: adjust everything up or down to match your headcount.

You'll have noticed that this book takes a make-your-own approach, on the principle that a store-bought jar of mayonnaise, say, is unlikely to taste as good or be as nutritious as the stuff you whip up at home. But given the level of dedication implicit in the tuma, especially when cooking it outside the region, you may as well buy some ready-made cassava water if you find it: there'll be no stigma attached.

TO FEED
2 PEOPLE
YOU WILL NEED

1 TAMBAQUI OR SEA BASS
AROUND 2.5 KG, CLEANED
AND WITHOUT THE HEAD,
SLICED IN HALF LENGTHWISE

1 HEAD GARLIC
(OR ABOUT 6 CLOVES),
FINELY CHOPPED

UP TO 5 KG OF
CASSAVA ROOT,
OR 2 LITRES
READY-MADE
CASSAVA WATER

2 ONIONS, FINELY CHOPPED

150 G CHOPPED CELERY,
SPRING ONIONS,
BASIL AND THYME

8 WHOLE GUYANESE WIRI WIRI
PEPPERS (OR OTHER CHILLIES)
AND 2 OF THE SAME, CUT
(Reduce the quantity if you prefer
your food less spicy.)

2 TBSP SALT

METHOD

1 Peel your cassava roots and soak them in water for a couple of hours at least. When the time's up, throw out the soaking water and keep the roots. Dry them well, grate them, then press or squeeze them hard to extract the juice. Leave this juice to stand in the fridge overnight: it will settle and develop starch. (The leftover cassava shavings can be baked into bread and served alongside the dish.)

2 Take the starchy cassava juice out of the fridge and boil it to remove any last trace of toxicity. As the juice boils, it will foam. Collect the foamy top layer until you've exhausted the liquid; reserve it in a container – this is your *kadakura*. (The thick residue left behind is known as *cassareep* – another popular ingredient in Guyanese cooking.)

3 Brush all sides of the two halves of fish with a little oil, salting generously and rubbing with some of the chopped herbs. Grill each fish half, in a large griddle pan or on the barbecue, until charred yet still moist.

4 Pour the *kadakura* into a fresh cooking pot, adding the onion, garlic, the rest of the herbs, the cut and whole peppers and a good pinch of salt. Bring back to the boil.

5 Cut the grilled fish into large chunks and place it in the *kadakura*. Simmer for a further 6–8 minutes and serve with bread or a starch of your choice.

Surubí

PSEUDOPLATYSTOMA

Surubí (*surubim* in Brazil) is a large spotted catfish of the Amazon basin, also present in South America's other great waterways. (For a broader discussion of catfish, see p. 36.) Several species of the genus – there are currently eight recognized ones – contribute significantly to the diet in Bolivian and Paraguayan lands.

What is now the Plurinational State of Bolivia lost its seaboard in the late nineteenth century. Paraguay never had one. These are South America's only two landlocked countries; access to rivers is critical in both. The one hosts the southern reaches of the Amazon River system; the other is shaped by the eponymous Paraguay, the Paraná and their tributaries. (The name *surubí* comes from Guaraní, Paraguay's Indigenous and co-official language.)

Here as elsewhere, with droughts increasingly frequent, climate change is causing environmental and economic stress. In September 2021, the Paraná, which carries most of Paraguay's international trade and supplies much of its electricity, dropped to its lowest level in nearly eight decades. When damming and overfishing across the region are factored in, the threat to riverine species – alongside more immediate concerns – can be acute.

KNOW YOUR FISH

The mild-flavoured spotted surubí (*surubí pintado, Pseudoplatystoma corruscans*), which our recipe originally features, is among the larger catfish, measuring up to 1.5 metres or more. It is a handsome beast – Brazilians call it *peixe rei*, or king fish – with long wavy whiskers and markings that evoke whimsically meandering ink drops. It is also overfished.

Aquaculture involving the spotted surubí has generally favoured hybridization with another species, *Pseudoplatystoma reticulatus*, resulting in individuals with a longer reproductive cycle. In some cases, poor controls have allowed hybrids back into the river system; these have been known to crossbreed back with one or the other of their parental wild species, jostling ecosystems and complicating species recognition. Given the tiered conundrum of status, availability and identification, we recommend using monkfish or flounder instead.

Fish casserole

CHUPÍN DE PESCADO

In Paraguay – but also in Argentina and Uruguay – *chupín* is derived from the Genovese *ciuppin*, a dish that is half-soup and half-stew, made with fish and leftover bread. Italy's regions abound in rustic recipes that put stale bread to good use, such as the Tuscan (and more widely central Italian) panzanella or *pappa al pomodoro*. Some of these were carried over to South America – in this instance, by Ligurian emigrants. The Paraguayan version uses potatoes.

Paraguay cheese (*queso Paraguay*) is a soft to semisoft curdled milk cheese. Equivalents should be easy to source in most countries, whether domestically produced or – as is most likely in parts of Asia – imported.

TO FEED
4 PEOPLE
YOU WILL NEED

4 MONKFISH FILLETS
(IN REPLACEMENT OF
SURUBÍ) OF AROUND
200 G EACH

4 TOMATOES, DICED

4 MEDIUM-SIZED
POTATOES, THICKLY SLICED

4 CLOVES GARLIC,
ROUGHLY CHOPPED

400 G CREAM

400 G PARAGUAY OR
OTHER CURDLED MILK
CHEESE, FINELY CUBED

3 ONIONS, THINLY SLICED
•
2 CAPSICUM, DICED

1 LARGE KNOB OF
BUTTER

METHOD

1 Grease a roasting tin with the butter. Line the pan with the potato slices, then rest the fish fillets on top. Season with salt and pepper.

2 Sprinkle the chopped garlic over the fish and potatoes, then layer the onion slices on top, then the capsicum and tomatoes. Make sure you neatly cover the surface, leaving no gaps.

3 Pour the cream and distribute the cubed cheese over the dish. Refrigerate for an hour to allow the cream to seep through and the flavours to meld.

4 Bake in the oven at around 160° C, for 40 minutes or so, until cooked through. (The exact timings and temperatures will depend on your oven.) Serve hot.

I'm glad at least one of you could make it.
Yes, unfortunately tambaqui and surubí were otherwise engaged. At least, I know that's the case for tambaqui. Surubí didn't get back. I'm slightly worried, in fact – we don't see much of surubí lately. Overfishing, you know…

I'm well aware, sadly. You, though, pirarucu, staged a comeback that's little short of spectacular.
I know. Goes to show what sound policies can do, eh? Especially when they're made with local people, rather than against them.

There's a notion that you're a lot like cod on the plate…
Perhaps. But we don't intersect. I'm purely a river creature.

… and that you were once a young fighter from the Nala nation, vain and evil-doing, who was punished by the gods.
If so, it was a long time ago. I wouldn't remember. I came close to extinction, as you suggested earlier. There was hardly anyone left to keep species memory alive. In any case, if I did anything wrong, I'm sorry. You know what they say about not revisiting the sins of the fathers.

Of course. Before we conclude, as tambaqui couldn't make it – would you like to say a few words on its behalf?
Actually, I have a message from tambaqui, let me read it to you. "Please send fruit and nuts."

I'm not sure I have any.
Well, it was worth a try.

THE INTERVIEW
RIVER FISH OF THE AMAZON BASIN

SHELLFISH

From the extravagant Tyrian purple pigment, secreted by the Murex snail, to the Judaic ban on fish devoid of fins and scales; from the glamour of pearls, the oyster's response to alien intrusions, to the vulgar co-option of mussels and clams as a proxy for female genitals; from legends about the immortality of lobster to nightmares about giant sea insects; and from aphrodisiac powers (bogus) to violent allergies (real) – for most of recorded human history, shellfish has concentrated fantasies, anxieties and interdicts. And that's because it's thrilling.

Unlike most of the fish described in this book, shellfish – it's not even really fish, but more on that below – is a mental voyage. Hardly anyone enjoys shellfish from birth: we learn to. Calling it an acquired taste might suggest we import our liking for shellfish in one fell swoop. We don't: we grow into shellfish as smokers grow into cigarettes. Shellfish is, of course, much less likely to kill us than cigarettes – though on occasion, tragically, it will. In other, thankfully exceptional instances, it can make us forget ourselves. Domoic acid, a marine biotoxin, has been known to lodge itself inside bivalves: once consumed by mammals, it causes amnesic shellfish poisoning (ASP), an irreversible form of memory loss.

Where fish may please us, shellfish dares us. Rather than offer, shellfish conjures – a marine squall, a shard in our heel, a lungful of brine as we first learned to swim. Out of a scallop shell, as myth would have it, Venus was born; and it is scallop shells, affixed to trees, that guide the pilgrim to Santiago de Compostela. Fertility and beauty. Vengeance and transcendence. A link to the out-there. If fish is the thing, shellfish is the metaphor. Oysters (Ostreidae) are rarely eaten to fill stomachs (even though in Japan, the practice of breading and deep-frying them – *kaki furai* – does bring them closer to fish fingers). They're eaten to evoke. We don't chew oysters. They transit our mouths like comets, setting off intimations of minerality. In Seamus Heaney's words,

"My palate hung with starlight:
As I tasted the salty Pleiades
Orion dipped his foot into the water."

Mucuslike, hard-plated, bug-eyed or antennaed: the fascination of shellfish admittedly encloses a kernel of repulsion. Molluscs may remind us of garden slugs or nasal secretions; crustaceans – of various crawlies. Indeed, as members of the *Tetraconata* clade, shrimp and lobster are genetically much closer to cockroaches than they are to fish. Possibly the rarest, most bizarrely shaped and costliest of all shellfish are the *percebes* found off the Iberian coast (*Pollicipes pollicipes*), which look like gnarled bouquets of dinosaur feet. Shellfish throws barriers in our way, physical and psychological. But think of the endorphins released by a strenuous run; of the meditative delights of abstract art; of the cerebral high born of breaking a code. The finest pleasures are those that don't come easy.

Here as elsewhere, effort correlates with value. Up until the nineteenth century in some cases, and across much of the world, the relative difficulty of harvesting shellfish led to its use as currency: shellfish was, one might argue, the bitcoin of its time. To this day, in Chinese and Japanese scripts, the character for "mollusc" (貝) also denotes money, or features as a radical in composite characters that signify a monetary transaction: its origins lie in representations of the cowrie shell.

Away from mercantile associations, shellfish keeps us going in unseen ways. Molluscs are filter feeders: they neutralize pollutants and control algal biomass. Their presence correlates with biodiversity hotspots. But there's more. In recent years, with climate change's scorching breath down our necks, we've woken up to the capacity of molluscs for carbon sequestration. Farming shellfish, some research suggests, could be a more effective form of climate action than planting trees, with comparatively minimal inputs by way of irrigation, food or fertilizer. An oyster, moreover, will remove – or "sink" – carbon permanently. Eating oysters, in other words, might be one of the best things you can do for the environment. (This is not, mind you, a licence to stop recycling your garbage.)

And yet, the present section doesn't contain oyster recipes. True, frying or grilling oysters can be rewarding. On p. 109 we suggest adding them to an all-in-one béchamel-based Chilean fish dish. Still, a raw oyster is unbeatable. You might be offered lemon juice, vinegar, tabasco, Worcestershire sauce, chopped shallots, ponzu dressing or whatnot: we recommend you shrug it all off. Unless it fits into a wider culinary endeavour, an oyster shouldn't be tamed, qualified or defanged. Like a glass of wine, it needs no adornment: it is an experiential closed loop, a gustatory Moebius strip. Shucking aside, scraping it gently off its shell is all you need to do.

HOMARUS

Contrary to enduring myth, lobsters are not immortal. They do die naturally, though they have no biological clock. In this, they are different from most living creatures: rather than age in a conventional sense, lobsters eventually succumb to the repeated, ever-more-exhausting effort to moult – that is, to shed and rebuild their formidable exoskeleton. In between dropping the old shell and growing a new one, their soft flesh is easy prey.

Mortality and such vulnerabilities aside, research points to lobsters' adaptability and resilience. The Woods Hole Oceanographic Institution, a non-profit body in Massachusetts, has found that where other sea creatures – corals among them – might corrode and dissolve, lobsters develop thicker shells in response to anthropogenic ocean acidification. (This is the process whereby the build-up of CO_2 from human activity poisons marine life.) One separate study in Denmark discovered that the submerged rigging of off-shore windfarms was a popular breeding ground for lobsters – an environmental double scorer.

Scientific and advocacy battles continue over whether lobsters feel pain. They're essentially insects, one side of the argument goes: they have no brain, no cortex; their thrashings when boiled alive (as they have been traditionally) are mere reflexes, not manifestations of suffering. There is, of course, as much literature to suggest the contrary. At the time of writing, New Zealand, Norway and Switzerland, all trendsetters for animal welfare, had banned the practice of scalding lobsters to death. The city of Parma in Italy, the nation's gastronomic heartland and home to the European Food Safety Authority (EFSA), has done so by municipal ordinance. Similar legislation is in the pipeline elsewhere.

Given all this and until we know better, we strongly recommend erring on the side of humanity: please stun or kill your lobster before cooking it, even if the law where you are doesn't say you must. Some jurisdictions also make it illegal to freeze a live lobster – all the more reason to get yours as fresh as possible.

KNOW
YOUR FISH

Lobster should be almost smell-free, with the kind of marine fragrance that would be called *vivifiante* in French – briskly rejuvenating, in the way of a sea breeze.

Since you're unlikely to be able to stun your lobster electrically, a quick kill is achieved by plunging a chef's knife into the base of the head, close to where it meets the "tail" (body), then swiftly bringing the knife down, splitting the head in half. Once cooked (watch out for overcooking lobster: it makes it disappointingly elastic), cut off the head. Turn the tail over, flatten it gently and press on the sides to break the shell and release the flesh. Now twist the claws off the knuckles. To get at the meat inside, whack each side of the claw two or three times with the knife. Wriggle the shell away, pulling carefully on the smaller pincer: it should come off with the tendon, leaving you with a whole, tender, sweet, claw-shaped piece of meat. The mess of lobster shells, together with any gooey leftovers, can be refrigerated for a day; use it to make shellfish stock or a bisque.

One word on langoustines (or scampi – *Nephrops Norvegicus*). These are sometimes presented as a smaller variety of lobster but are more accurately described as a type of crab. Their nutritional profile reads like a supplement label – vitamin B12, selenium, potassium, copper, iodine, and all with very little fat content. Split them down the back and roast them in a hot oven for a couple of minutes, or until the flesh turns from translucent to just white, and no longer. Or steam or grill them. Serve them with a splash of olive oil and a sprinkle of salt flakes, or maybe a couple of capers and a dash of mandarin juice. Langoustines, that said, will be special-occasion stuff for most of us: the meat-to-weight ratio is barely 20 percent, against 50 percent or so for lobster. Their flavour is highly enticing, but their price is, alas, unenticingly high.

Nutrition facts

SPINY LOBSTER, FLESH, RAW
per 100 grams

ENERGY (kcal)	91
PROTEIN (g)	19
CALCIUM (Ca) (Mg)	37.8
IRON (Fe) (Mg)	1.1
ZINC (Zn) (Mg)	3.1
IODINE (I) (µg)	24
SELENIUM (Se) (µg)	45
VITAMIN A (RETINOL) (µg)	3
VITAMIN D3 (µg)	0
VITAMIN B12 (µg)	2.2
OMEGA-3 PUFAS (g)	0.13 *
EPA (g)	0.06 *
DHA (g)	0.07 *

* These values are from a similar species: American lobster, flesh, raw

Lobster pasta with grated tomato sauce

The overwhelming majority of lobsters consumed in the United States of America – and that's a great many – are harvested off the state of Maine. In 2021, the haul hit a record value of over USD 700 million, as consumer demand rebounded after the first year of the COVID-19 pandemic. By weight, the catch held more or less steady at around 45 000 tonnes – but that in itself is five times as much as fisherfolk were bringing in 40 years ago. The warming waters off the northeastern seaboard are thought to have been propitious for lobsters (although there are fears a cliff edge could be near). The thinning out of cod, lobster's main predator, may further explain the bonanza of recent decades.

Our lobster recipe comes courtesy of Carla Lalli Music, a US television chef and cookery writer. In her book _Where Cooking Begins_, she draws on her Italian heritage to create dishes that are stylishly uncomplicated. While there may be a touch of dexterity involved in shelling the lobster, the greater challenge here is arguably not to overboil the pasta – something that almost invariably, and not without some justification, Italians think pretty much everyone else is guilty of.

We've adapted Lalli Music's recipe to reuse the lobster's steaming water to cook the spaghetti, for added flavour. If you like a touch of spice, we also suggest enlivening the sauce with peperoncino chilli – or, if meat is no object, with a fiery smear of nduja, the Calabrian hot salami paste.

TO FEED **2 PEOPLE** YOU WILL NEED

I LIVE LOBSTER

3 LARGE BEEFSTEAK TOMATOES

250 G DRY SPAGHETTI OR LINGUINE

I HANDFUL TORN BASIL LEAVES

4 CLOVES GARLIC, SLICED

I KNOB BUTTER (OPTIONAL)

SALT AND PEPPER

METHOD

1 Kill the lobster by splitting its head as previously described. Pour 5 cm salted water into a large saucepan and bring it to a boil. Now place the lobster in the water or in a steamer above it, and cover the pot. Bring back to a boil and cook for 6 minutes if the lobster is sitting in the water, or 8 minutes if it's in the steamer, until it's not quite cooked through.

2 Remove the lobster, crack the shell and retrieve the meat, taking care to keep the claws whole. Cut the tail into bite-sized pieces. Cover with cling film and reserve.

3 Supplement the water left after cooking the lobster and add more salt. Bring to a boil again.

4 Cut a slice off the base of each beefsteak tomato, then, holding the tomatoes by the stem, push them through a grater (the side with the large perforations) into a large bowl. Discard the skins.

5 In a large frying pan, heat up some olive oil (and any butter if using) with the chopped garlic over medium heat, until the garlic is fragrant. If using nduja or any chilli, add that too at this point. Tip the grated tomato in, and season with salt and pepper. Simmer for 10 minutes or so, allowing the sauce to thicken and reduce. Throw in the basil and remove from the heat.

6 Now place the pasta in the pot of boiling water. Once the water is boiling again, keep it going for 3 minutes less than it says on the packet – it should be very *al dente*, supple but at the threshold of edible, as it will continue to cook in the sauce.

7 Return the sauce to a medium heat and transfer the spaghetti into it, adding a little of the pasta water if needed. Cook the spaghetti for a couple of minutes until *al dente*, then add the lobster in, warming it through and no more. Ensure that both diners get a claw with their portion when serving. You may want to add a drizzle of olive oil and a fresh grind of pepper. For added crunch – this, again, is our own twist – you could sprinkle freshly fried breadcrumbs on top.

173

Mussels and clams

MYTILIDAE, VENERIDAE

Mussels and clams – species of which are so varied and numerous that placing them in a single family remains controversial – are often lumped together, despite being quite distinct. Both are bivalves, as are oysters and scallops. They secrete their own shells out of calcium carbonate, expanding it to accommodate their growing selves. (The human equivalent would see us enlarging our childhood rooms at will, as we bloom into adolescence and adulthood.) Mussels, which tend to cluster on marine cliffs and river shores, are easily grown on longlines, in tubular nets or, more anciently, on ocean rafts. Almost the entire global production comes from aquaculture, much of it from China. Most mussels, however, are consumed in Europe, and some are produced there too.

Unlike mussels, clams are infaunal animals, that is, living in sediments. They too may be cultured, but to a lesser extent, and with generally longer farming cycles. When harvested wild, clams need digging out individually from the sandy seabed: they are, all told, considerably pricier than mussels.

In much of the world including parts of the Mediterranean, clam-digging was customarily women's business, often entailing hardscrabble wages. The sector remains intensely gendered in Tunisia, a country that counts some 4 000 women digging at nearly 20 sites. Here, around the coastal cities of Gabès and Sfax, FAO has been working with the clam-diggers' association to cut out exploitative middlemen and link the women directly to importers. The price paid to the diggers is higher and guaranteed; they also receive a premium for only picking the larger clams, which makes for a more sustainable supply. Training in handling and food safety is part of the package.

The main clam markets all have their preferences, driven by access to native species. In the United States of America, these will often be the hard-shelled quahogs of Rhode Island, famed as the stuff of chowder; in Japan, the colourful Manila clams, used in a clear, cleansing broth called *ushio-jiru*; in Spain, razor clams, which pair felicitously with warm olive oil and garlic, but equally well with jamón and toasted almonds; in France, the *palourdes* with their yellow glow, slick with butter and parsley; and in Italy, the darker *vongole*, tossed with spaghetti in a reduction of garlic, chilli and white wine, the whole enriched by the addition – wholly optional – of *guanciale*, or pig's cheek bacon. Swine and clams also click happily in *porco à Alentejana*, the Portuguese laurel-scented land-and-sea stew.

Mussels are, by comparison, a straightforward, economical affair. While arguably less varied or versatile, they're no less enjoyable than clams, frequently fleshier and sporting a springy chewiness. Mussels are also richer in iron and lower on sodium, should you be watching your salt intake. Their even plumpness is a repeatable joy – and the fun of using one valve to scoop the mollusc out of the other will get the children hooked.

Nutrition facts

MEDITERRANEAN MUSSEL, FARMED, FLESH, RAW
per 100 grams

ENERGY (kcal)	65
PROTEIN (g)	8.3
CALCIUM (Ca) (Mg)	59.5
IRON (Fe) (Mg)	2.5
ZINC (Zn) (Mg)	1.9
IODINE (I) (μg)	140
SELENIUM (Se) (μg)	49
VITAMIN A (RETINOL) (μg)	68
VITAMIN D3 (μg)	0
VITAMIN B12 (μg)	14.2
OMEGA-3 PUFAS (g)	0.40
EPA (g)	0.21
DHA (g)	0.16

KNOW YOUR FISH

You will need to soak your clams in cold salted water for up to four hours in advance, during which time they should expunge any sand they hold. Bubbles will form in the water; change it once or twice. If a lot of sand is released, throw each clam lightly against the side of the kitchen sink to force out the last grains. When you've done this, scrub the clams – still in cold water – if they appear to need it. We offer a recipe that marries clams with fish, crab and other crustaceans: you can find it on p. 191.

Since mussels are overwhelmingly sourced from aquaculture, they will generally come prerinsed, with no need for soaking. Still, running them under cold water, and maybe giving them a quick scrub, will remove any extant sand or grit. Finally, you may have to yank off the "beard," technically known as byssus. (These wispy strings, once prized as sea silk for use in textiles, are secreted by the animal to bind it to its habitat.) As with all bivalves, discard any open mussels, or those that don't open up when exposed to heat. And mind that you don't keep them long: mussels spoil faster than clams.

Our two mussel recipes come from the northern and southern hemispheres. Both are frank comfort food, as anything involving mussels should be, but differently so. One is firmly in the continental Atlantic tradition. The other melds Indigenous influences with modern US and Japanese ones.

Mussels

MOULES À LA BELGE / MOSSELEN OP Z'N BELGISCH

Moules-frites (mussels and chips) may well be the most reductive shorthand for Belgian-ness. This doesn't make it any less delicious or evocative – of briny gales at Ostend or Zeebrugge, of piers and carousels, of a heartwarmingly tacky fry-up kiosk on a foggy Saturday night. So intimate is Belgium's identification with the dish – the average citizen consumes 4 kilograms per year, twenty times the global average; on National Day in July, giant outdoor *moules-frites* dinners are laid in cities across the country – that hardly anyone cares to remember where everyone's favourite mussels come from, namely across the border, in the Dutch province of Zealand.

Je suis parti vers ma destinée
Mais voilà qu'une odeur de bière
De frites et de moules marinière
M'attire dans un estaminet...

"I set off to meet my calling
When a sudden whiff of beer
Of fries and of moules marinière
To a tavern sent me crawling"

sang the punk chansonnier Arno, who was raised a Fleming in Ostend but lived his artistic life in French. In federalized, bilingual Belgium, *moules-frites* are a cross-community value, whether the mussels are braised in ale, in a curry sauce or – as in our recipe – with wine, garden vegetables and herbs.

For the *frites*, you will no doubt have your own twist on the basic process. Remember also that in Belgium, a slather of mayonnaise comes close to a legal requirement. We tell you how to make it on p. 129.

TO FEED
2 PEOPLE
YOU WILL NEED

2 CARROTS

1 KG MUSSELS, CLEANED

2 LEEKS

1 ONION

OLIVE OIL FOR COOKING

2 CLOVES GARLIC, SMASHED

1 BOUQUET GARNI
(BUNDLE OF AROMATIC HERBS):
parsley, bay leaf (laurel) and thyme,
tied together with a piece of string

1 GLASS WATER

1 GLASS DRY WHITE WINE

SALT AND PEPPER

METHOD

1 Pour 2 tbsp olive oil and the cloves of garlic into a deep pan. In Belgium, this would traditionally be a black enamel saucepan. Place the pan over a low flame.

2 Dice up the vegetables – this is known as a *mirepoix* – and season them with salt and pepper. Add them to the pan, together with the cleaned mussels.

3 Add the wine and water to the pan and bring to a vigorous boil. Cover the pan and cook until the mussels have opened.

4 Scoop out the mussels into deep plates and spoon over the cooking liquid. Serve with the *frites* and mayonnaise or aioli.

Green–lipped mussel fritters

New Zealand is the land of *kaimoana* – the Maori word for seafood, integral to the Indigenous culture of this historically seafaring nation. One friend recalls a childhood indulgence in *mātaitai*, or shellfish gathering, as she'd chase *tuatua* clams, which squirt water in self-defence and burrow away at astonishing speed, or else the palm-sized *toheroa*, with their meaty, protruding tongues and gamey flavour. *Toheroa* are now almost extinct and strictly protected. Poaching them is subject to blistering fines, in a move echoing (belatedly from a conservation perspective) ancestral practices that combined ritual and stock-management concerns: in traditional Maori harvesting, no shellfish could be opened while people were in the water; only one species could be gathered at a time; and during certain periods, no picking could take place at all.

Even so, New Zealand is still, overall, replete with bivalves. The country counts more than 20 types of mussels, including the endemic green-lipped variety (*Perna canaliculus*). These are large mussels, easily twice the size found elsewhere, rimmed with a green so vivid as to be almost fluorescent. Cultured in large quantities since the 1970s, they're New Zealand's top seafood export, shipped under the "Greenshell" trademark as far as the United States of America, Spain and the Republic of Korea. Their nutritional value – they're full of protein, low in fat and brimming with omega-3 acids, vitamin B12 and selenium – also sees them processed into food supplements.

Mussel fritters are a popular dish with Maori and Pākehā (European) New Zealanders alike. To the extent that they're egg-and-milk pancakes with savoury ingredients baked in, they channel both American brunch classics and Japanese *okonomiyaki* waffles. You can use either Greenshells or your nearest available variety for our recipe: they'll be chopped up, so their pretty lips will be immaterial in this case.

TO FEED
2 PEOPLE
YOU WILL NEED

1.5 KG GREEN-LIPPED MUSSELS

2 EGGS, SEPARATED

50 G SELF-RAISING FLOUR

50 ML MILK

1 LARGE HANDFUL PARSLEY, FINELY CHOPPED

SALT AND PEPPER

VEGETABLE OIL FOR FRYING

METHOD

1 Bring 2-3 cm of water to the boil in a large saucepan. Add the mussels in batches and cook covered for 2 minutes, or until the mussels have just opened. Remove the mussels to a large bowl, discarding any unopened ones, and place the bowl on ice.

2 When the mussels are cool enough to handle, shell them and chop them up finely, manually or in a food processor.

3 Separately, combine the egg yolks, flour and milk to make a batter. Mix in the chopped mussels, together with the sliced spring onions and chopped parsley. Season with salt and pepper.

4 Now mix the egg whites until stiff and fold them into the batter. Heat the oil in a skillet (1 tbsp should do for a couple of rounds), ladle in each fritter and fry till crispy brown. Serve with a lemony or mustardy salad.

Conch

STROMBUS GIGAS

The exquisite shell of the conch (queen conch to US readers) will be familiar to many. We'll have seen it in childhood, perhaps, gracing an elderly relative's mantlepiece or display cabinet. We might have been handed it, gingerly, with an injunction to hold it to our ear, so we could "hear the sea in it". This notion of a live marine broadcast with its rolling waves is, of course, fanciful – the swirly cavity of the shell simply amplifies ambient noise – but also indicative of conch's rich semiotics. The mollusc's function as a proto-loudspeaker, with solemn or quasi-mystical attributes, appears to echo through a number of traditions. In Hindu religious practice, the *shankha* shell is used as a ceremonial trumpet. In William Golding's *Lord of the Flies*, only the boys who hold the conch may speak: through it flows the voice of authority.

Materially, the conch shell is extremely tough, its structure potentially replicable through 3D printing to devise near-unbreakable helmets and body armour. Three-layered, it features what one research team at the Massachusetts Institute of Technology (MIT) has described as a "zigzag matrix": any cracks, the team told *MIT News*, would be forced to "go through a kind of maze".

A little lost in all the talk about conch is the living creature itself – the *actual* conch, as opposed to the shell whose biological rationale is, after all, to shield the animal within. This is a sea snail, indigenous to the Florida Keys and the Caribbean. (Natives of Key West endearingly refer to each other as "Conchs".) Its meat goes into burgers, soups, stews and curries. It is occasionally seen raw, or as ceviche in Panama. An ancestral source of protein for the area's Arawak and Carib people, it has a sweet, delicately oceanic flavour, and a consistency that's somewhat cartilaginous. If you enjoy textured foods, this chewy slipperiness is a pleasure in itself. Bear in mind that as a rule, the larger the conch, the more rubbery: cook it too long, or not enough, and you may render it inedible.

Nutrition facts

CONCH SHELLS, FLESH, RAW

per 100 grams

ENERGY (kcal)	104
PROTEIN (g)	23.3
CALCIUM (Ca) (Mg)	271.2
IRON (Fe) (Mg)	3
ZINC (Zn) (Mg)	4.4

KNOW YOUR FISH

You may find your conch to buy fresh or frozen.

If you're getting it preshelled – this is the default case with queen conch, which tends to be shucked on collection and cleaned at packing plants – look out for meat that is rose-yellow in colour. Depending on its size and toughness, you may wish to slice your conch very thin, or pound it to tenderize it, or else run it through a meat grinder. Marinating it in lime or kiwi juice is another option.

Smaller varieties of conch may be sold whole: you can cook these in their shells, then tease out the cooked meat using a shellfish fork. Discard any grey knobbly parts, which contain the digestive gland and suchlike. Finally, cut off and throw out the operculum, the hard disk that propels the live conch along the sea floor.

Vincentian conch soup

LAMBIE SOUSE

Queen conch is far from all-you-can-eat stuff. Overfishing led Florida to ban commercial and recreational harvesting in the mid-1980s. The 1990s saw CITES, the convention that regulates trade in endangered species, list the animal; the body has occasionally shut down production in some countries, pending the adoption of sustainable management plans. Culturing of conch – for education and tourism more than commercial purposes – has only been practised at one farm in the Turks and Caicos Islands. Devastated by the hurricanes of 2017, that facility has closed indefinitely.

Caribbean nations these days have a host of controls in place: these involve bans on immature conch, period closures and geographic restrictions. The archipelago of Saint Vincent and the Grenadines numbers some 45 licensed conch – also called "lambie" – fishers, operating from long open boats between May and August (when the season for lobster, another main source of income, is closed).

The haul, brought up by divers, makes up two-thirds of the country's fisheries exports. Diving, however, is hazardous work, and FAO has been looking to develop safer capture techniques around the Caribbean. No easy task, this: in fishing subcultures, diving remains powerfully connoted as a masculine exploit.

Our Saint Vincentian recipe is flavoured with *chadon beni*, French Creole for *chardon béni*, or "holy thistle". Sometimes known as culantro, this is a cousin of coriander, but sturdier and more potent. Outside the region, just use coriander and dial up the quantity – a handful will do perfectly. The *lambie souse* is best made in a pressure cooker.

800–900 G CONCH MEAT

TO FEED
4 PEOPLE
YOU WILL NEED

1 ONION, CHOPPED

1 CUCUMBER, CHOPPED

1 TBSP CHOPPED PARSLEY

1 TBSP CHOPPED CELERY

1 TBSP CHADON BENI OR 1 BUNCH
CORIANDER

1 SQUEEZE LIME

SALT, PEPPER AND
(OPTIONAL) HOT PEPPER

METHOD

1 Wash the conch meat in lemon juice or vinegar, then place it in a pressure cooker with the *chadon beni*. Cook for 30 minutes. Do not salt at this point.

2 Place all the chopped vegetables and herbs in a separate container (or soup dish). Remove the conch from the pressure cooker and, taking care not to burn yourself, cut it into bite-sized pieces. Add it to the other chopped ingredients.

3 Pour the conch stock from the cooker into the soup container, mixing all ingredients well. Season with salt, pepper and hot pepper if using. Serve warm.

Prawns

DENDROBRANCHIATA, CARIDEA

Prawns are entry-level shellfish. Firm and smooth, often seen with just their graceful tail left on, they're somehow normalized, at a pincer's length from the messy insect-ness of other crustaceans. Children like them; almost everyone does. Giant Indo-Pacific tiger prawns, smoky with barbecue juices. Tiny European prawns sold in a pint glass, to be gobbled like peanuts. *Amaebi* from northern Japanese waters, eaten raw, their flavour sweet, their mouthfeel creamy. (These cold-loving prawns switch from male to female in the course of their life: they're at their sweetest when young, before the spontaneous gender reassignment occurs.) Prawns in a paella, or a ceviche, or a jambalaya, or served with discs of radish over rye bread on a Swedish *Midsommar* evening. There's no getting away from Caridea, to give the creatures their formal name, and nor would we want to.

But there's a cost to our taste for them. The industry faces regular flak for environmentally or ethically questionable practices. Prawn trawling in the wild clearly sweeps up a plethora of other, untargeted creatures. This is, in all fairness, a scourge of the wider global fisheries sector: it's estimated that over half a million marine mammals alone, not to mention turtles and other vulnerable species, die as bycatch every year. In 2021, FAO laid down guidelines for preventing and reducing the phenomenon. These comprise a mix of institutional approaches and technical solutions such as acoustic deterrents and vessel monitoring systems.

But the charge sheet doesn't stop there. For farmed prawns, routine eyestalk ablation involves cutting off females' eyes to hasten sexual maturity and spawning. While the exact neurotransmission process has yet to be understood, the practice has been branded cruel and stress-inducing; it has also been shown to triple mortality rates and damage eggs' resistance to disease. The list continues with toxic spills and the destruction of mangroves to make room for prawn farms in Asia, where production is heavily concentrated. (Other parts of the world are catching up. Preliminary data for 2021 suggests Ecuador may have unseated India as the top exporter. In early 2022, the war in Ukraine forced the closure of Europe's largest prawn farm, after just four months of operation.)

Consumers can help minimize damage – and partly assuage cruelty concerns – by buying certified prawns, including farmed organic ones. Most certification schemes focus on environmental sustainability, but animal welfare is emerging as an additional parameter, and so are working conditions and pay in the industry. The Global Sustainable Seafood Initiative (GSSI), for example, derives benchmarking standards from FAO's Code of Conduct for Responsible Fisheries and other internationally -negotiated instruments. The sector has, all told, upped its game over the last decade or so. Mangrove destruction has not only decreased, but been partly reversed. And biosecurity, another of FAO's areas of action, is more robust. On Thai prawn farms, for instance, the arrival of sophisticated water-exchange systems, the use of pond liners, and the introduction of biofloc – a technology that purifies sludge and excrement and turns it back into food – have greatly reduced viral outbreaks.

Nutrition facts

RIVER PRAWNS, FLESH, RAW
per 100 grams

ENERGY (kcal)	87
PROTEIN (g)	16.6
CALCIUM (Ca) (Mg)	43.7
IRON (Fe) (Mg)	1.1
ZINC (Zn) (Mg)	2.5
IODINE (I) (μg)	120
SELENIUM (Se) (μg)	26
VITAMIN A (RETINOL) (μg)	10
VITAMIN D3 (μg)	0
VITAMIN B12 (μg)	[7]
OMEGA-3 PUFAS (g)	0.09
DHA (g)	0.08

KNOW YOUR FISH

Depending on their size, you may want to devein your prawns – that is, remove the digestive tract, a thin black tube that runs the length of the creature. You can do this by cutting lengthwise into the back, grabbing the upper end of the tube, and peeling it off. The procedure will make for a classier result, but unless the prawns are quite large, it won't make much of a difference on the palate.

When it comes to innards, predilections vary. Where western tastes run to scrubbed-down prawn flesh, savvy Asian consumers will know to squeeze the head of the animal for its immensely flavourful contents. If you haven't tried it, you'll probably wish you had. Collect two of three tablespoonfuls of "shrimp butter" and fry it for a few seconds in hot oil, until it turns a bright brick colour. Spread it on toast, or use it to season anything from pasta to chicken soup to Pad Thai. Or pour it over plain long rice, with a frank tear of coriander.

Stir-fry prawns with Longjing tea leaves

LONGJING XIAREN, 龙井虾仁

From a small player in the late 1970s, China has catapulted into the world's top league of freshwater prawn producers. But even this level of growth has been outpaced by internal demand. Urbanization and rising incomes have sent domestic consumption of prawns soaring. As much as half of the Chinese market (of combined wild and cultured, marine and freshwater species) is now supplied by imports from Ecuador, another rising prawn power; a further quarter comes from India.

Our recipe marries China's enthusiasm for freshwater prawns – historically a southern delicacy of more limited currency – with the prestige of Longjing tea. This pale-leafed, mildly astringent, roasted green tea is harvested around the West Lake, the famed setting for the city of Hangzhou. Its fragrance is often described as evocative of chestnuts: you may also find it sold under the Dragon Well name. Legend connects this prawn dish to a visit

to Hangzhou by the eighteenth-century Qianlong Emperor, said to have been travelling in disguise: a local innkeeper, we're told, serendipitously mistook the Longjing leaves for green onions while feeding the sovereign.

Chinese cooks are fond of "velveting" their prawns by dousing them, as we do here, in egg whites, cornflour and salt. The

process provides a pleasing touch of crunch; it also insulates the prawns from the searing wok heat and keeps them moist.

Shaoxing wine is used for depth of flavour. If you've ever cooked Chinese food, you'll likely have it in your pantry already. If not, it'll be readily available wherever you are in the world.

TO FEED
2 PEOPLE
YOU WILL NEED

1 TBSP LONGJING
(DRAGON WELL) TEA
LEAVES

400 G SEA OR RIVER PRAWNS,
not too large and preferably fresh, fully shelled,
heads off (and deveined if desired)

2 EGG WHITES

2 TBSP VEGETABLE
COOKING OIL

60 CL SHAOXING
COOKING WINE

2 TBSP CORNFLOUR

METHOD

1 Pour 2 tbsp Shaoxing wine and the egg whites into a bowl, then tip in the prawns and coat them in the mix. Add the cornflour and a pinch of salt and coat the prawns further – you want the coating neither lumpy nor completely homogenous: occasional floury protrusions will add crispy texture. Cover with cling film and refrigerate for half an hour.

2 With the prawns in the fridge, make a cup of Longjing tea. Don't strain it – the leaves must stay in. The water should be below boiling temperature to preserve the fragrance. Let the brew stand a while.

3 Take the prawns out of the fridge. In a wok, warm up the oil and toss in the prawns. Flash-fry them, stirring quickly, until they are semi-translucent white ("like jade"). This could be a matter of seconds. Remove them to a colander to strain any dripping oil.

4 Now clean the wok and return it to the flame. Add the tea, leaves included, and the remaining Shaoxing wine. Simmer this down a little, then put in the prawns and braise them briefly. Serve them glossy with the reduced liquid.

Crab

BRACHYURA

Crabs cross the saltwater-freshwater divide. Between them, the thousands of known species make up a fifth of all crustaceans harvested around the world, whether from capture or aquaculture – about 1.5 million tonnes. Some crabs are so small as to be barely visible: they generally hold little interest as human food, though the pea crab (*Pinotheres pisum*), a tiny oyster parasite, is considered a delicacy in some quarters. At the opposite extreme, two tall men could fit between the splayed legs of the Japanese spider crab (*Macrocheira kaempferi*).

Along that spectrum we find the five or so main commercial species. At the largest, rarest and costliest end sits the spiny red king crab (also known as Alaskan king or Kamchatka), from the frigid waters of the Bering Sea. Then comes the smoother-shelled snow crab, more widely distributed around northern oceans; the smaller, white-purple Dungeness, present along North America's Pacific shoreline; the stone crab, plump from munching on conch and other sea snails off Florida, which sheds and regenerates its claws at will: fishers will just tear off one claw and throw the animal back into the sea to regrow it; and the more economical rock crab, particularly present in the eastern Atlantic, from the Arctic to Mauritania.

Intriguingly, while the question of whether crabs feel pain is much the same as for lobster, crabs don't trigger nearly the same amount of empathy. Could this be because, edibility aside, crabs carry representational associations with disease? In around 400 before our era, the man credited with laying the bases of medicine as we know it, Hippocrates, mysteriously chose to name cancer after the crab: *karkinos* in Greek. (Analogies between the pattern of malignant cells and the shape of crabs are likely a postfactum conceit.) In some modern languages, the two terms remain identical. And even as we scoff our crabs with abandon, we continue to saddle them with morbid projections – for example, in referring to their gills, unappetizing but harmless otherwise, as "dead man's fingers".

As we went to press, the war in Ukraine was reverberating through the world crab trade, with Western sanctions spurring a reconfiguration of the Russian Federation's exports. The country holds nearly 95 percent of the global quota for red king and snow crab: in the previous year, it pulled in USD 2.4 billion in revenue. Much of the haul used to go frozen to the United States of America, the European Union and allies. There are now signs that more will go live to China instead.

Nutrition facts

KING CRABS/STONE CRAB, FLESH, RAW

per 100 grams

ENERGY (kcal)	69
PROTEIN (g)	15.8
CALCIUM (Ca) (Mg)	103
IRON (Fe) (Mg)	0.6
ZINC (Zn) (Mg)	6
SELENIUM (Se) (μg)	36
VITAMIN A (RETINOL) (μg)	7
VITAMIN D3 (μg)	0
VITAMIN B12 (μg)	9
OMEGA-3 PUFAS (g)	0.19
EPA (g)	0.13
DHA (g)	0.05

KNOW YOUR FISH

Most crabs prefer cool (or even gelid) waters, and so tend to feature in the cuisines of cold or temperate lands. But their popularity has been spreading in warmer latitudes. In India, the mangrove-dwelling mud crab is no longer the preserve of coastal communities: it has been picked up by urban chefs. The Singaporean chilli crab, laced with a hot, eggy tomato sauce, seems poised for global fame. In Thailand, soft shell crab – harvested when the animal is moulting, before its exoskeleton firms up – is a classic, rolled in seasoned flour, then deep-fried and served with a spicy dip. A hop away in French-influenced Viet Nam, the dish may be accessorized with lettuce leaves to wrap the crab in, plus masses of fresh dill.

Gills and guts aside (discard those), crab offers both white meat (from the claws) and brown (from the body). Which is better? Well, it rather depends: the debate mirrors the one about chicken breast vs chicken leg. We'd rather have both, frankly. And if you chance upon a female crab with roe, we suggest you also try that soft, friable matter. Much like the coral of scallops – often omitted in the belief that anything but the blandest stuff will scare away consumers – it delights as a textural and chromatic counterpoint.

Parihuela

From Peru, one of the world's fishing heavyweights and gastronomic powerhouses, comes another all-in-one: the spicy parihuela seafood soup, which combines fish and crustaceans – not just crab, but mussels and clams and prawns and sea snails. That, however, is what you might call a *parihuela* royale – fit for a banquet, maybe, but hardly the kind of thing you'd whip up for a dinner with friends. It's also the case that the list of ingredients speaks to the peculiar bounty of Peru's coastline: elsewhere, securing all the species might prove both a logistical headache and a blow to the wallet.

Ultimately, once you have a fish, a crab and a handful of clams, everything else is an optional luxury. We recommend that you kill the crab first, by splitting it in half: as well as being the more humane route, this will ensure that the internal crab juices leach out during cooking and further flavour the stock. On how to purge the clams, see "Mussels and clams" on p. 174.

There is, admittedly, a further difficulty involved in this dish, in that it uses the distinctly Peruvian ingredient *chicha de jora*. This is a refreshing, marginally alcoholic (1 percent to 3 percent) drink obtained from fermented corn. *Chicherías* are found in Peruvian cities, but also, less formally, in the Andean villages of the south – often simple holes in the wall, advertising themselves by means of a bamboo pole that sticks out over the road, with a roll of colourful fabric or plastic bags wrapped around its end. If there is no Peruvian bodega where you live, and no *chicha* can be sourced online, try replacing it with apple cider: you may need to reduce it a little beforehand, as *chicha* is thicker. In Türkiye and parts of the southern Balkans, *boza* – the fermented barley drink, vaguely fizzy and low in alcohol – would make an exciting alternative.

190

TO FEED
4 PEOPLE
YOU WILL NEED

4 FISH FILLETS
(cod, halibut, seabass or another white fish of your choice), plus the head or other cut-offs for stock

2 MEDIUM-SIZED CRABS, CLEANED AND SPLIT DOWN THE MIDDLE

I RED AND I YELLOW CHILLI PEPPER, TO TASTE, DESEEDED AND CHOPPED

I DOZEN PRAWNS OR SNAILS (OPTIONAL).
If using shrimp, cut off the heads and save them.

I HANDFUL OF CLAMS OR MUSSELS, OR BOTH, CLEANED BUT WHOLE

200 ML CHICHA DE JORA OR REDUCED APPLE CIDER

I ONION, CHOPPED FINE

I TSP SMOKED RED PEPPER PASTE*
•
SALT AND PEPPER COOKING OIL
(canola, peanut or sunflower) for frying

I TSP FRESH GRATED GINGER

2 CLOVES GARLIC, CHOPPED LARGE

I BUNCH CORIANDER (CILANTRO) FOR GARNISHING

METHOD

*(Peruvians use *ají panca*, a sweet, mildly spicy product; substitute with chipotle powder or paste if needed, perhaps mixing it with a little sugar, or with smoked paprika.)

1 Place the fish head, together with the prawn heads if using, plus any fishy cut-offs in a large saucepan. Cover with water, season and bring to a boil. Turn down the flame and let simmer until the stock has reduced by half. Skim off any scum and taste regularly to ensure the broth has enough flavour. When it does, remove from the flame and allow to cool, then strain.

2 In another saucepan, sweat the onion, garlic, chilli and grated ginger in oil, stirring in the red pepper paste or paprika, to make the *sofrito* base. Keep the flame low and stir to avoid burning, until the mix is soft and fragrant.

3 Now add the *chicha de jora*, cider or other substitute. Turn up the flame and let the liquid reduce somewhat.

4 Add the halved crabs to the *sofrito* and *chicha* mix, gently coating them in it, then toss in the clams or mussels and any snails. Cover with a lid to concentrate the heat and allow the bivalves to open.

5 At this point, add the fish fillets, spooning some of the red sauce over them, and pour the strained fish stock over the lot. Bring back to a gentle boil and simmer for 5 minutes.

6 Let the *parihuela* cool for a few minutes, during which the flavours will further intermingle. Spoon it into bowls, ensuring everyone gets something of everything. Sprinkle some fresh coriander over each bowl and serve with chunks of bread or a starch of your choice.

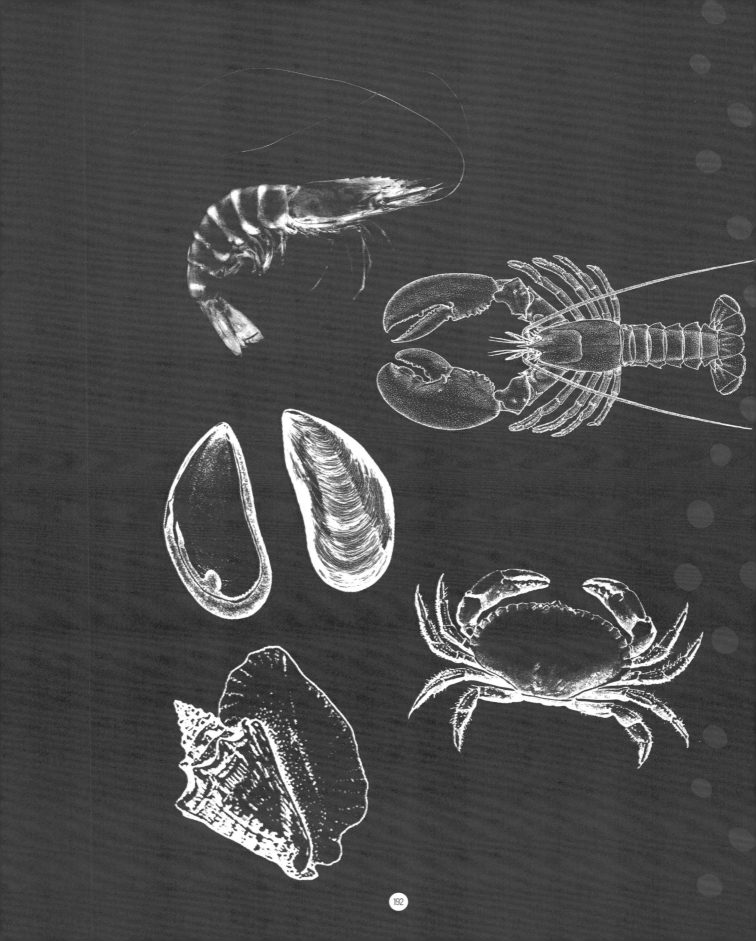

It's great that you've all turned up.
[CONCH] What was that? I'm a little hard of hearing.

Well, no wonder with that tough shell you've got. Why don't you come out, conch?
[CONCH] I always thought humans were meant to listen to me, not the other way round. And anyway, it's too dangerous to come out. You've got crab and lobster there.
[LOBSTER] What? Me? Us? How dare it? I'll pinch the little—

Now, now, let's all calm down. Mind your language, please, and don't talk over each other. Lobster, as you're the biggest and most expensive one here, you should set an example. Kindly keep your claws down – we don't encourage threatening behaviour.
[LOBSTER] Humph.

Everybody should have a voice. And you too, crab. I hope you didn't take my comment to mean you should step aside.
[CRAB] But I must. It's what I do.

I see. Then let me turn to you, prawns. I understand you have a message for our readers.
[PRAWNS] Yes, thank you for the opportunity. Our message is this: we know many of you only want us for our bodies. Which is terribly demeaning. We have excellent heads. Try them.

Good advice. But it occurs to me we haven't heard from the clams...
...

Nothing?
[MUSSELS] Clams don't speak. And we're pretty taciturn too. But we have another way of expressing ourselves.

Ah? And what is it?
[MUSSELS] Just listen.

To what? I can't hear anything!
[MUSSELS] Shhh... We're cleaning the sea for you...

THE INTERVIEW
SHELLFISH

GREAT CHEFS
COOKING
GREAT FISH

ÁNGEL LEÓN

MICHELIN-STARRED EMBRACE OF BYCATCH

Ángel León is Spain's *Chef del Mar*, the "Sea Chef" celebrated for fashioning wonders out of the Mediterranean's humblest, most overlooked produce: here are sardines as you've never had them before, and plankton as you've never had it... well, ever. This frugality-themed inventiveness has earned León three Michelin stars and the Green Star for Aponiente, his restaurant at El Puerto de Santa María close to Cádiz, and a further star for his newer venture nearby, Alevante. Both menus evince a half-rigorous, half-playful quest for a form of marine autarky – a longing to substitute the sea for the land in all things, or almost. Why use butter to thicken a sauce, when you can use fish eyes? Why mess about with egg white to clarify a consommé, when micro-algae will do? And why, while we're at it, throw out anything at all? If the sea made it, then you can eat it.

WATCH VIDEO

Fish mortadella
©Manuel Montero/
Handafilms

ÁNGEL LEÓN'S FISH MORTADELLA

For our book, León has chosen a recipe that is both a feat of lateral thinking and simplicity incarnate. He has looked to mortadella, the Italian heat-cured luncheon meat sausage known to Americans as baloney, and recreated it with *morralla*. This is "no-name" fish – bycatch consisting of flathead grey mullet and suchlike, generally discarded or used at best for stock or surimi crab sticks. At the opposite (but equally fish-minded) end of Europe, Norwegians speak of *ufisk*, literally "un-fish". *Ufisk* is fish too cheap or ugly to serve in its own right, the kind of stuff you might even get for free – if the fishmonger can be bothered to stock it.

León, by contrast, rescues the fillets from the bycatch and grinds them up, then adds mortadella spices – typically a finely crushed mixture of salt, white pepper, garlic powder and pistachio nuts – which he works thoroughly into the ground fish flesh. (The pistachios can be left chunkier, or even whole.) For the next step, León stuffs this matter into a synthetic skin – in effect, a narrow transparent plastic bag – to obtain a fat cylinder which he seals shut at both ends.

The "sausage" is then placed in a hot water bath (or steam bath) on the hob, at exactly 82 degrees Celsius for two hours. When the time's up, León plunges his creation straight into an ice-filled bowl. As the protein cools, the fish sausage acquires a mortadella-like texture and cut.

Slice your fish mortadella thin and have it on crusty white bread.

MALEK LABIDI

TRAILBLAZER WITH NATIONAL FOLLOWING

For Malek Labidi, landing an economics degree in Paris was a concession to expectations, an academic baseline from which to follow her true path: that of professional kitchens. Having come up through the Institut Paul Bocuse (a period she describes as a gruelling but exhilarating three years), she cut her teeth with Alain Ducasse at the Plaza Athénée hotel in Paris before returning home to Tunis to open Le Bô M. That restaurant broke new ground by offering a daily changing menu, in a conservative market hooked on tried-and-tested fixtures. Several years on, Labidi has acquired a national following through catering State occasions and fronting TV cookery programmes. An advocate of fresh produce and healthy diets, she updates her homeland's versatile cuisine – looking both out, across the Mediterranean, and inland, towards sun-soaked gardens and orchards.

MALEK LABIDI'S SALT-BAKED SEABASS IN SAUCE VIERGE

Labidi has cooked for us a whole seabass immersed in a salt crust. The fish should be gutted, but the scales must stay on. Half the pleasure of eating seabass is its moist mouthfeel, almost a flavour in itself: the salt seals in this moisture while seasoning the fish through its skin. The sauce vierge, which involves dressing tomatoes and herbs in lukewarm olive oil, coats the seabass like a scented balm.

As well as a whole fish, Labidi uses a large quantity of coarse sea salt – four kilograms or so. She also uses an orange for zest; a lemon for both zest and juice; a red onion; long green peppers; a large bunch of fresh dill, stems included; dry thyme; a sprig of rosemary; fresh ginger; juniper berries; and olive oil.

First, Labidi pre-heats the oven to 200 degrees Celsius. Meanwhile, in a large deep bowl, she mixes the salt with some chopped dill, thyme, juniper berries that she crushes lightly, and grated ginger and lemon zest. (If this is too much to mix in one go, do so in batches; also, keep some dill for later.) Labidi then lays a thick layer of aromatic salt in the bottom of an oven tray, places the fish on top, and covers it completely with the rest of the mixed salt. The tray goes in the hot oven for 20 minutes.

To make the sauce vierge, Labidi warms olive oil gently in a saucepan for three or four minutes and takes if off the flame. Into this saucepan she cuts pieces of orange and lemon zest (pith removed), a couple of slices of ginger and the sprig of rosemary, and leaves these to infuse the oil. The onion, tomatoes and green peppers are chopped finely and placed in a separate bowl. She squeezes some lemon juice over the vegetables, seasons them with with salt and pepper, and strains the oil over them.

The seabass now comes out of the oven. Labidi cracks and removes most of the salt crust, then peels away and discards the salt-encrusted skin. The pale, glistening flesh emerges: she collects it and plates it up. Over it, she drapes the sauce vierge, with its fresh palette of reds and greens.

WATCH VIDEO

Salt-baked seabass
in sauce vierge
©Slim Bouguerra

MEGHA KOHLI

SEASONAL DELIGHTS WITH MILLENNIAL APPEAL

Megha Kohli's dedication to food is no late conversion. Aged 4, she recalls riding on her father's shoulders through the spice markets of Old Delhi. At six, she was making cakes in her grandmother's kitchen. A quarter of a century later, she admittedly cooks in a more gown-up manner, but with that early buoyancy intact. Kohli has now transferred to Café Mez and The Wine Company in the upscale Delhi suburb of Gurgaon. But her reputation was established at the more central Lavaash by Saby. At that Armenian-Indian eatery, she turned culinary heads by (among other things) baking coconut prawns inside blossoming onions. Her many online followers seem to lap up Kohli's style: intimate yet crowd-pleasing, and – as seen here in her refusal to peel her mangoes and her doubling down on the fruit's skins – infused with a witty millennial touch.

MEGHA KOHLI'S RAW MANGO FISH CURRY

Kohli uses mango three-ways in this dish: powdered (*amchur*); slices cooked from raw, skin-on; and extra raw peels for what she calls an "additional punch". The recipe calls for green mangoes: you don't want a sweet wet mess. The skins are powerfully nutritious, full of fibre and antioxidants. Wash them well or go organic to minimize the presence of pesticides.

For the fish, Kohli uses boal (*Wallago attu*), a variety of catfish, which she marinates in advance with lemon juice, salt and turmeric. Any other freshwater fish that doesn't flake excessively will do just as well. The lemon and turmeric help remove any whiff of geosmin, the organic compound that tends to accumulate in freshwater fish. Kohli recommends a "darne" cut, that is, thick crosscut pieces sliced through the bone. The recipe also calls for curry leaves; black mustard and cumin seeds; more turmeric; coconut milk; dried red chillies; several garlic cloves, crushed; and chilli powder. The cooking is done in ghee, the clarified butter almost universally used for frying in India. (Alternatively, you can use a neutral cooking oil with a high burning point – preferably not olive oil or butter, which are too intensely flavoured and tend to smoke heavily.)

Kohli pours ghee into a non-stick pan and adds the mustard and cumin seeds over a medium flame. When the seeds start to sizzle and pop, she puts in the sliced raw mango, the extra mango skins and the powdered ingredients – *amchur*, turmeric and chilli. She softens the mango for a couple of minutes, then tips in the coconut milk and cooks the sauce gently for another minute. The pieces of fish go in next, to simmer until done. (Depending on their thickness, this could take five to seven minutes or so.) Taste the sauce, season with salt as desired, and adjust the consistency with a little water or fish stock if needed.

The final step is the tempering, or *tadka* – the process whereby spices are blasted to the next level of taste by flash-frying them in hot fat. Kohli heats a little more ghee until it starts to smoke: she throws in the curry leaves, the crushed cloves of garlic, more mustard seeds and the dried whole chillies. She sautés the lot for half a minute and drizzles it on the fish and sauce. The curry is served over steamed rice.

WATCH VIDEO

Raw mango
fish curry
©Saumya Gupta/
Shiva Kant Vyas

RODRIGO PACHECO

FAO GOODWILL AMBASSADOR: LOCAL PRODUCE, GLOBAL FLAIR

With its seaside location, Bocavaldivia might suggest a beach outpost of some temple of fine dining from Quito or Guayaquil. It is, in fact, one of a kind – the gastronomic wing of a comprehensive site-specific project. Here, Chef Rodrigo Pacheco and his team have regenerated a stretch of vacant land into an "edible forest" – a microcosmic food system of biodiverse farming and agro-ecological practices. In his wall-less kitchen, Pacheco marries Indigenous ingredients to seafood of fine character. Ostentation is off the menu: the edibles are near-abstract swirls of colour against unglazed black clay. But behind the spare aesthetic and short supply chain lurks great complexity of process. This is a coastal and urbane re-reading of Amazonian cuisine: local food as global discourse and sustainability template.

RODRIGO PACHECO'S PRAWN *NEA PIARAKA* SOUP WITH CASSAVA CRISP

For our recipe, Pacheco sources prawns fresh from one of the fishing crews in nearby Puerto Cayo – several large, visually appealing crustaceans, plus a couple of smaller ones for flavour; homegrown cassava – both the fresh root and fermented flesh, as well as the leaves; *neapia*, a dark, spicy and smoky condiment in the form of a dense jam made with cassava starch, from which the soup takes its name; the natural, mildly peppery orange pigment *achiote* (also called *annatto*) – Pacheco grinds this into a marinade, still textured from the berries of the plant, and seasons it with salt and lime juice; garlic; chives; and finally culantro – the long-leafed, punchier cousin of coriander, which many Ecuadoreans know as *chillangua* and natives of the Caribbean as *chadon beni*.

The cassava root is peeled and grated to obtain a fistful of wet fibre, which is squeezed hard. The whitish, semi-translucent juice is collected into a bowl; the remaining dry fibre is set aside. Also into the bowl go the culantro, cassava leaves and chives; the chopped-up smaller prawns; half a teaspoon of *neapia*; and a small clove of garlic, lightly smashed – all of which will infuse the cassava juice.

Pacheco then peels and deveins the larger prawns, leaving the heads and tails on, and brushes them with the *achiote* marinade. Next, he turns to the dry cassava fibre, left after squeezing out the juice, and sifts it through a colander into a layer of coarse flour: he bakes this into a slice of crispy Indigenous bread (*casabe*). Separately, the dollop of fermented cassava flesh is cooked to make a diminutive tortilla. The prawns, turned a rich red from the *achiote* marinade, go under the grill.

Pacheco now assembles the dish. In a deep earthenware plate, he places the tortilla. Over it, he ladles the strained infused cassava broth, fragrant with herbal and marine notes. The grilled prawns are then lowered in to rest on their tails. More cassava leaves are added as vertical green accents. The final element is the *casabe* crisp: balanced over the stack, it signals the dish's Indigenous inspiration.

WATCH VIDEO

Prawn *nea piaraka*
with cassava crisp
©Ángel Lucio C.

SANDRO SERVA
MAURIZIO SERVA

RENAISSANCE OF THE FAMILY RESTAURANT

The phrase "family restaurant" in Italy will likely conjure a pleasant but predictably folksy experience. Nine times out of ten, the reality will be just that. At La Trota dal '63, the reality is anything *but* that. The sixty-year-old venue sits in a hamlet among wooded hills and crystal lakes. Here, brothers Sandro and Maurizio Serva and their respective sons, Michele and Amedeo, serve up food of inordinate audacity. La Trota offers nothing but freshwater fish – the only European restaurant at this level of excellence to do so. Sandro and Maurizio work the kitchen: self-taught, they have elevated what was a pit-stop trattoria to a two-Michelin star establishment. Michele and Amedeo are front-of-house: they contribute their wine knowledge and design nous. The fish served here hails from around the corner; the experience belongs somewhere altogether less homely.

WATCH VIDEO

THE SERVA BROTHERS' EEL WITH A WATERCRESS INFUSION AND KIWI

Maurizio Serva cooks for us a new entry on this year's menu. (If you fancy a go, start prepping a couple of days in advance.) The treatment preserves the toothsomeness of eel flesh while offsetting its fatty intensity. As well as the fish, Serva uses its backbone; a bunch of watercress, whose presence attests to the purity of local waters; fresh dandelion; sugar; orange, lemon and lime; a kiwi; and spices.

Serva whizzes the cress in a blender with ice water to create a grassy, peppery juice. In this juice, he places a portion of eel, which he leaves to marinate in the fridge for 48 hours. The backbone of the fish he cleans completely and dries out in the oven at 80 degrees Celsius for at least 12 hours. When the bone is dry, he fries it in very hot oil, allows it to cool and blends it into a powder. (This powder, he explains, will convey the quintessence of the eel's flavour.)

Next, Serva simulates a dandelion "honey". In a small saucepan, he makes a syrup with water and brown sugar, to which he adds cinnamon, star anise and a clove, together with grated lemon, lime and orange zest. With the syrup off the flame, he puts in a handful of dandelion flowers and leaves, adds a good squeeze of lemon juice and leaves the thick liquid to infuse overnight. The next morning, he strains it: the dandelion honey is ready.

Serva now turns to the kiwi. He heats up a skillet, trims the fruit, cuts it in half lengthwise and sautés it. He then sprinkles the cooked half of the kiwi with brown sugar and caramelizes it with a blow torch, before tackling the portion of eel. This he oven-roasts over bay leaves, then glazes it with the dandelion honey and blitzes it under the grill.

It's time to plate up. The eel and kiwi (cut side down) are arranged side by side. Over them goes a dusting of the fish bone powder. Next to the plate, in a refreshing herbal counterpoint, Serva places a cold glass of the filtered cress marinade.

Eel with a watercress
infusion and kiwi
©Dario Coronetta

References

Arno, Caussimon, J.-R. & Ferré, L. 1995. Comme à Ostende. In: *À la française* [musical recording]. France, Delabel.

Chang, Y.L.K., Feunteun, E., Miyazawa, Y. & Tsukamoto, K. 2020. New clues on the Atlantic eel's spawning behavior and area: the MidAtlantic Ridge hypothesis. *Scientific Reports*, 10(15981). https://doi.org/10.1038/s41598-020-72916-5

Cucherousset, J., Boulêtreau, S., Azémar, F., Compin, A., Guillaume, M. & Santoul, F. 2012. "Freshwater Killer Whales": Beaching Behavior of an Alien Fish to Hunt Land Birds. *PLoS ONE* 7(12).

Dante Alighieri. 1995. Purgatorio, Canto XXIV. In: A. Mandelbaum (trans.) *The Divine Comedy*. London, Penguin Random House.

Davidson, Alan. 1972. *Mediterranean Seafood*. London, Penguin.

Forster, E.M. 1926. Notes on the English Character. *Atlantic Monthly*, January.

François, B. 2021. *Éloquence de la sardine. Incroyables histoires du monde sous-marin*. Paris, J'ai lu.

Gray, J.H. 1988. The Flying Fish. In: *The Poems of John Gray*. Greensboro, North Carolina, USA, ELT Press.

Heaney, S. 2001. Oysters. In: *Field Work*. London, Faber and Faber.

Hogan, Z.S., Moyle, P.B., May, B., Vander Zanden, M.J. & Baird, I.G. 2004. The Imperiled Giants of the Mekong. *American Scientist* 92(3).

Jiwani, S. 2019. The shrinking pomfret of suburban Mumbai. In: *People's Archive of Rural India*. Mumbai, India. Cited 4 August 2022. ruralindiaonline.org/en/articles/the-shrinking-pomfret-of-suburban-mumbai/

Karateke, H.T. 2013. *Evliyā Çelebī's Journey from Bursa to the Dardanelles and Edirne*. Fifth Book of the Seyāḥatnāme, Volume 7. Leiden, Netherlands, Brill.

Kurlansky, M. 1999. *Cod: A Biography of the Fish that Changed the World*. London, Penguin Random House.

Martel, Y. 2011. *The Life of Pi*. Toronto, Canada, Knopf.

Oceana. 2019. Casting a Wider Net: More Action Needed to Stop Seafood Fraud in the United States. In: *Oceana*. Washington, DC. Cited 4 August 2022. https://usa.oceana.org/reports/casting-wider-net-more-action-needed-stop-seafood-fraud-united-states/

Pliny the Elder. 1950-1991. Book IX. In: H. Rakham, W.H.S. Jones & D.E. Eichholz (trans.) *Natural History*. Cambridge, Massachusetts, USA, Harvard University Press.

Ries, J.B., Cohen, A.L. & McCorkle, D.C. 2009. Marine calcifiers exhibit mixed responses to CO_2-induced ocean acidification. *Geology* 37(9).

Risso, A. 1810. *Ichthyologie de Nice, ou histoire naturelle des poissons du département des Alpes-Maritimes*. Paris, F. Schoell.

Sadovy de Mitcheson, Y.J., Linardich, C., Barreiros, J.P., Ralph, G.M., Aguilar-Perera, A., Afonso, P., Erisman, B.E., et al. 2020. Valuable but vulnerable: Over-fishing and under-management continue to threaten groupers, so what now? *Marine Policy*, Vol. 116.

Schmidt, J. 1923. The *Breeding Places* of the *Eel*. In: *Philosophical Transactions* of the *Royal Society* of *London. Series B, Containing papers* of a biological character, Vol. 211.

Stuntz, G.W., Patterson III, W.F., Powers, S.P., Cowan, Jr., J.H., Rooker, J.R., Ahrens, R.A., Boswell, K., et al. 2021. Estimating the Absolute Abundance of Age-2+ Red Snapper (*Lutjanus campechanus*) in the U.S. Gulf of Mexico ("Great Red Snapper Count report"). Mississippi-Alabama Sea Grant Consortium, NOAA Sea Grant.

Most information about the conservation status of species comes from the International Union for Conservation of Nature (IUCN). For specific nutrition data not present in the FAO/INFOODS Global Food Composition Database for Fish and Shellfish (uFiSh), we have drawn on the Aquatic Food Composition Database, and on resources of the Norwegian Institute of Marine Research (IMR) and the United States Department of Agriculture (USDA).

Acknowledgements

Many cooks – most of them amateur, some professional – have contributed recipes for this book. (Some recipes have been adapted.) Others, cooks or otherwise, have shared personal memories and cultural insights. A few of the contributors have been credited in the text; most have not. Their support and advice have been essential. We would like to thank them here:

Adeola Akinrinlola	Ani Grigoryan	Piotr Ogar
Marcelo Barcellos	Simeon Hall Jr.	Hugo Podesta
Tuğçe Basaran	Dejan Karapeev	Claudio Quiroz
Karisha Chakma	Frances Kennedy	Lina Pohl Alfaro
Francesco Di Bona	Alicia Lavia	Mairam Sarieva
Fatimatou Diallo	Nara Lee	Henri Schoenmakers
Jamon Edwards	Jeanne Marion-Landais	Shara Seelall
Dora Egri	Moseka Mokwa	Kyōko Shibuta
Guzal Fayzieva	Leshan Monrose	Alizèta Tapsoba
Izzat Fejdi	Valeria Navas Castillo	Juan Tinoco
Rosa Fonseca	Mark Nelson	Gustavo Vilner
Amy Collé Gaye	Vidyawatie Nidhansing	Orisia Williams
Sharine Gomez	Emely Nyakumuse	

Our thanks also go to the Missions of Myanmar and Sweden to the United Nations agencies in Rome.